US MILITARY ENCYCLOPEDIAS

# THE WAR ENCYCLOPEDIA

BY CYNTHIA KENNEDY HENZEL

Encyclopedias

An Imprint of Abdo Reference

abdobooks.com

# TABLE OF CONTENTS

# ARMED CONFLICT IN US HISTORY

The United States has fought many wars in its history. The first was the American Revolution (1775–1783). Victory in this war established the United States as an independent nation. The United States later fought several conflicts to gain land. These included the Mexican-American War (1846–1848) and the Philippine-American War (1899–1902). These wars expanded US territory to the Pacific coast and beyond. The American Civil War (1861–1865) started when some Southern states tried to break away from the country. The war kept the United States together and led to the abolition of slavery.

In the first half of the 1900s, the United States became involved in worldwide conflicts. In World War I (1914–1918), the country went to the aid of Western Europe against Germany. In World War II (1939–1945), US forces fought Germany again. At the same time, they battled Imperial Japan in the Pacific. By the end of this massive conflict, the United States was a dominant nation economically, politically, and culturally.

After World War II, the United States faced off against the Soviet Union in a struggle known as the Cold War. The two sides didn't fight each other directly, but they supported different

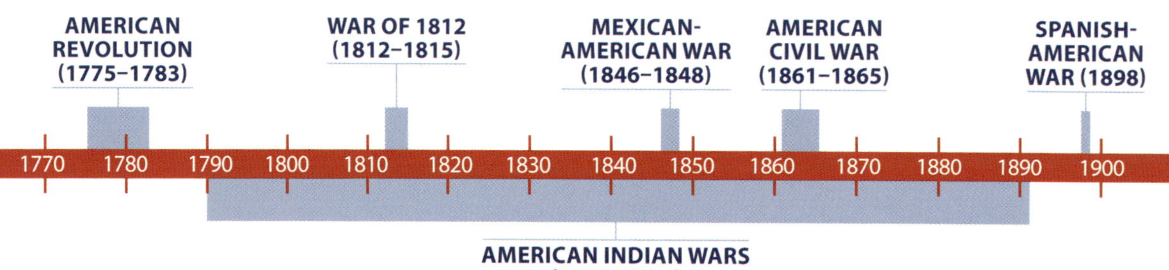

sides in conflicts across the globe. These wars included the Korean War (1950–1953) and the Vietnam War (1954–1975). The United States sought to prevent the spread of communism, which was the economic and political structure of the Soviet Union.

The United States later became involved in conflicts in the Middle East and Central Asia. In the Persian Gulf War (1990–1991), the United States defended Kuwait from an invasion by Iraq. In the Afghanistan War (2001–2021), US forces targeted those who harbored the terrorists behind the September 11, 2001, terrorist attacks that took place on US soil. And in the Iraq War (2003–2011), the United States removed dictator Saddam Hussein from power amid accusations that he was developing weapons of mass destruction.

The United States' many wars are an important part of the nation's history. They have shaped the country and are connected to cultural, scientific, and political changes. Understanding these wars, why they were fought, and how they were waged is important to understanding the story of the United States.

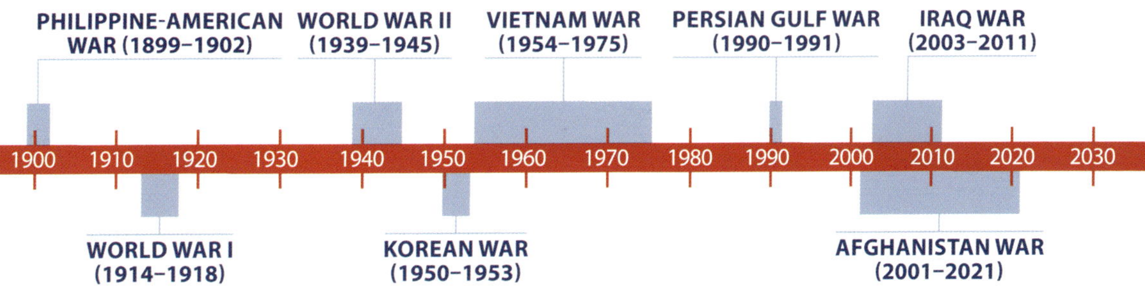

# AMERICAN REVOLUTION (1775–1783)

In the 1600s and 1700s, Great Britain established colonies along the eastern coast of North America. By the mid-1700s, some colonists were becoming upset at what they believed was an unfair relationship with Great Britain. After defeating

The part of the Seven Years War fought in North America was known as the French and Indian War. Great Britain and France each joined forces with different American Indian groups.

Great Britain produced stamps for enforcing the Stamp Act, but no colonial newspapers agreed to use them. Some simply stopped publishing instead of using the stamps.

France in the Seven Years War (1756–1763), Britain needed money. The British Parliament, which made laws and oversaw the government in Great Britain, passed the Sugar Act in 1764. It taxed sugar coming into the colonies. Then, in March 1765, Parliament passed the Stamp Act. The Stamp Act required all legal and commercial documents, including newspapers, to be stamped by the government, showing that a tax on them had been paid.

Many colonists opposed these acts. The colonists were British citizens and paid taxes to Great Britain. Yet they had no representative in the British Parliament. There was no one in the British government to speak on their behalf.

At first, the colonists refused to use the stamps. In the summer of 1765, Samuel Adams and a group called the Sons of Liberty began intimidating the tax collectors. Their motto was "no taxation without representation." In one incident, they broke into the home of Massachusetts's lieutenant governor,

Samuel Adams became a leading figure pushing for American independence.

Colonists rioted in New York City to protest the Stamp Act, which they saw as unfair.

Thomas Hutchinson. When Hutchinson refused to speak against the Stamp Act, rioters damaged and looted his home.

Leaders from nine colonies met at the Stamp Act Congress in October 1765. They wrote Parliament demanding the repeal of the Stamp Act. Parliament said no. As a result, colonists refused to buy British goods. This hurt British merchants. Parliament repealed the Stamp Act in 1766. At the same time, it passed the Declaratory Act. The act restated that the government had the power to tax people anywhere in the British empire for any reason.

A colonial engraving showed British troops arriving in Boston, Massachusetts, in 1768.

In June 1767, Parliament passed the Townshend Acts. These acts placed taxes on items imported into the colonies, such as glass, lead, paint, paper, and tea. But the colonists again resisted the taxes, and in 1770 Parliament repealed most of them, except for a tax on tea.

On the same day the Townshend Acts were repealed, March 5, 1770, a crowd of people protested the taxes in Boston, Massachusetts. The protesters faced off against British soldiers sent to protect the British authorities. One soldier fired into the crowd, and several others followed. In what was called the Boston Massacre, the soldiers killed five colonists and wounded six. A jury found six of the eight soldiers not guilty.

Colonist Paul Revere created a widely distributed engraving of the Boston Massacre.

# THE BOSTON TEA PARTY

On December 16, 1773, the Sons of Liberty demanded that a shipment of tea leave Boston Harbor to avoid having to pay the British taxes on it. Thomas Hutchinson, who was now the governor of Massachusetts, refused. The colonists disguised themselves as American Indians and boarded the ships. They dumped the tea into Boston Harbor. The event became known as the Boston Tea Party.

The British responded harshly. The Boston Port Bill closed Boston Harbor. The Massachusetts Government Act replaced the colony's elected council with an appointed council, taking away the right of the people to choose representation. The military governor, General Thomas Gage, received more power. Citizens could not hold town meetings more than

About 60 colonists participated in the Boston Tea Party.

The First Continental Congress met at Carpenter's Hall in Philadelphia.

once a year. The Administration of Justice Act allowed British officials accused of crimes to have trials outside the colonies, giving those officials more favorable conditions. Finally, the Quartering Act allowed British soldiers to be housed in people's homes. The colonists called the new laws the Intolerable Acts.

In response, 12 colonies sent 56 representatives to a meeting called the First Continental Congress. It began in Philadelphia, Pennsylvania, in September 1774. The Congress agreed to boycott British goods. The colonies would also stop sending goods to Britain until the Intolerable Acts were repealed. In addition, the Congress listed the rights of colonists. One was the right "to life, liberty, and property." A second right was to participate in government. But at this time, most colonists did not want a separate country. They wanted to avoid war.

# WAR BEGINS

By 1775, King George III and Parliament were tired of protests from the colonies. They ordered General Gage to stop potential challenges to British authority. He decided to send troops from Boston to destroy colonial weapons in Concord, Massachusetts. On April 18, Paul Revere and William Dawes rode on horseback to warn the colonial militia, which favored independence, that the British were coming. The militia prepared to defend Lexington, which was between Boston and Concord.

The British soldiers arrived at Lexington the next day. They ordered the

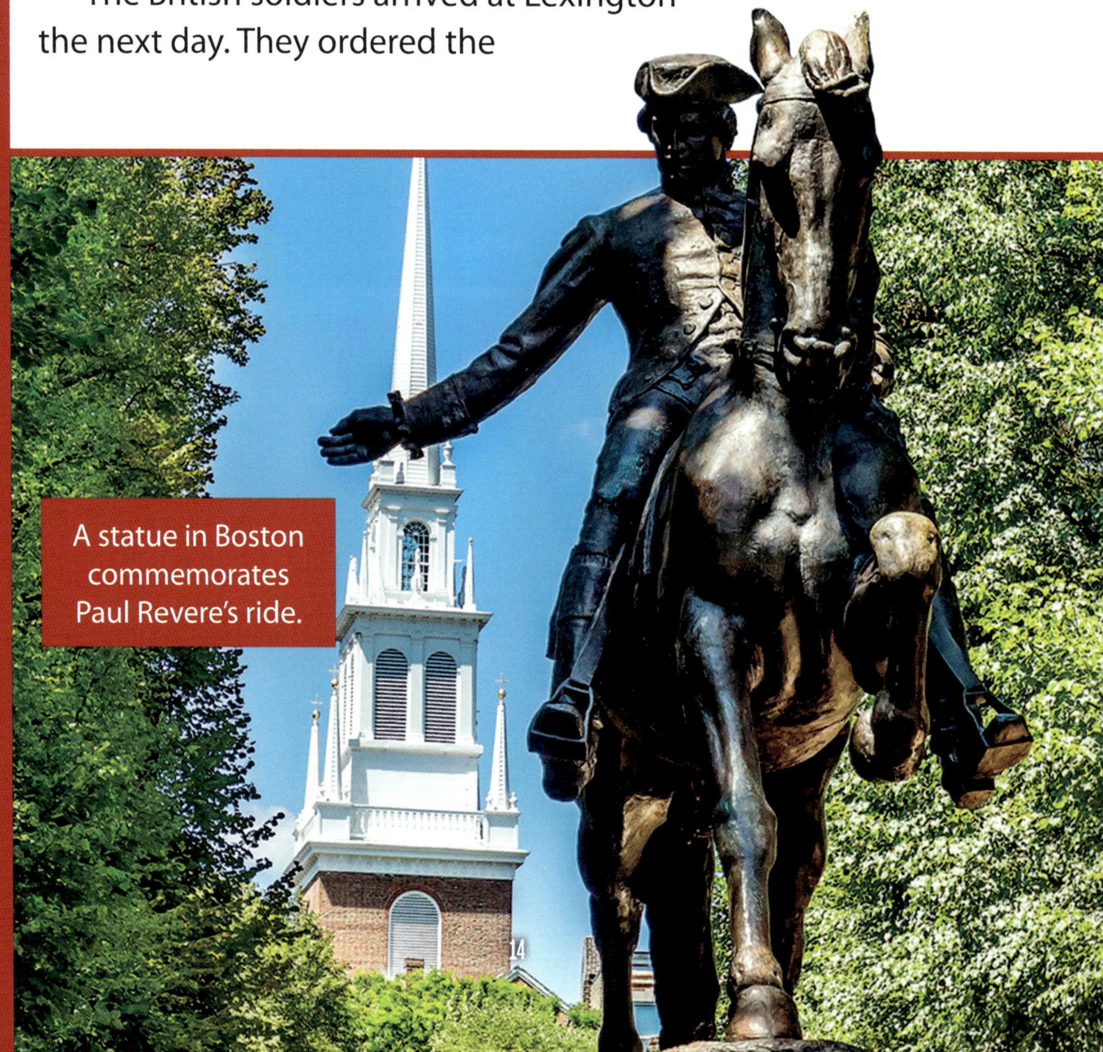

A statue in Boston commemorates Paul Revere's ride.

Modern reenactors show people what the fighting at Lexington may have looked like.

militia to leave. Shots were fired. No one knows which side shot first. The British killed eight colonists and then marched to Concord. However, most of the colonial supplies there had already been hidden or destroyed.

Fort Ticonderoga overlooks a narrow stretch of water at the far southern end of Lake Champlain.

As the British troops marched back to Boston, colonists hid along the roads and fired at them. In all, the British lost 273 soldiers and the colonists lost 95. The American Revolution had begun.

The colonists had their first major victory at Fort Ticonderoga in New York on May 10, 1775. Ethan Allen and his Green Mountain Boys militia rowed across Lake Champlain and captured the fort in one night. This kept the British from sending supplies into Canada for a northern attack. It also gave the colonists a supply of weapons.

In May 1775, the Second Continental Congress met. It appointed George Washington as commander of the

Continental Army. In June, the Congress agreed that the colonies must become free states. Thomas Jefferson wrote the first draft of a document called the Declaration of Independence. The members of Congress approved it on July 4, 1776. The Declaration explained the rights that people have. It listed the ways that King George III had violated those rights. The document stated that all people have a right to "life, liberty, and the pursuit of happiness." Finally, it proclaimed that "these United colonies are, and of the right ought to be, free and independent states."

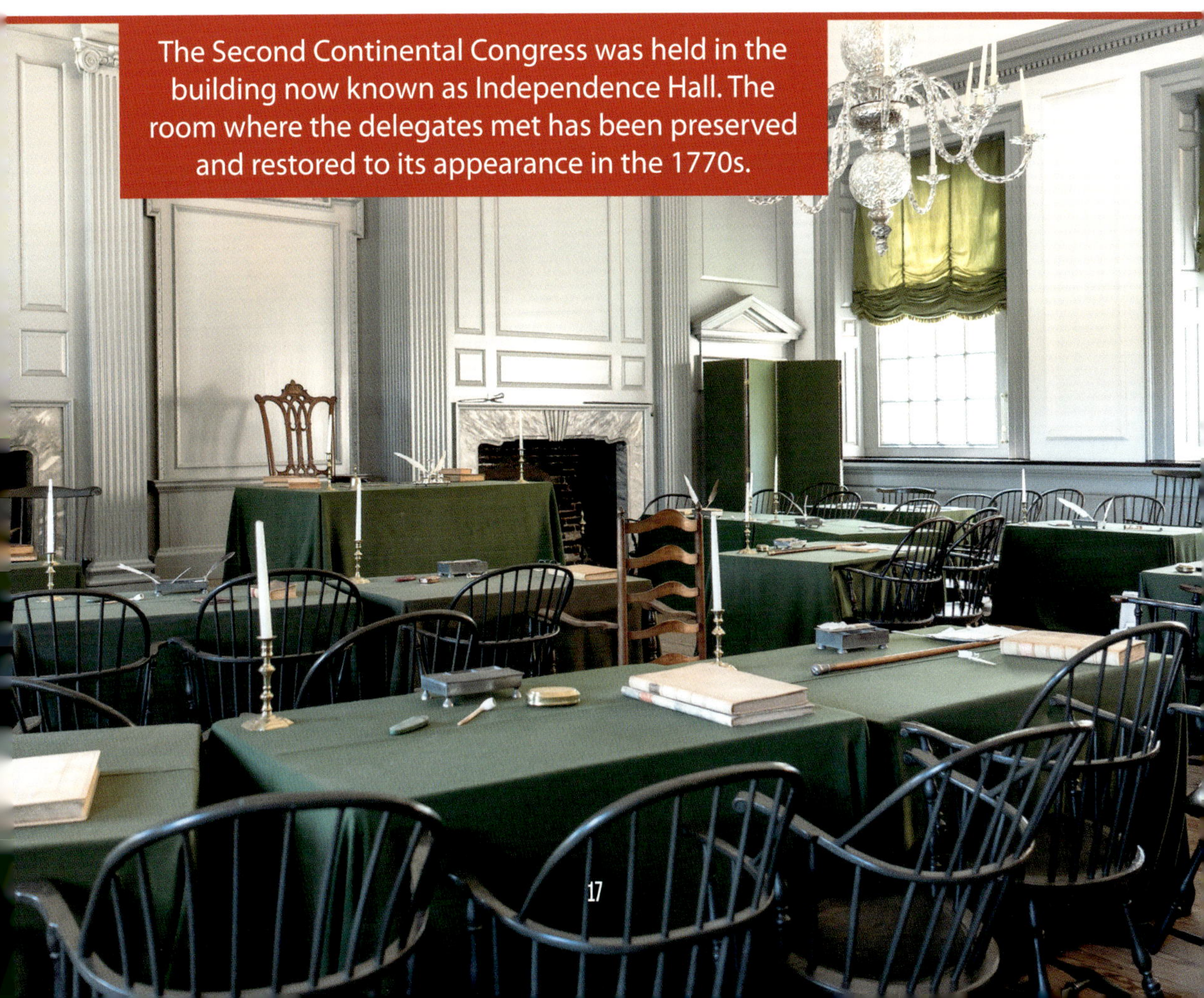

The Second Continental Congress was held in the building now known as Independence Hall. The room where the delegates met has been preserved and restored to its appearance in the 1770s.

## THE BATTLE FOR INDEPENDENCE

As the Second Continental Congress met, the two sides fought a major battle. It was called the Battle of Bunker Hill. It was fought on two hills in Charlestown, Massachusetts, called

Bunker and Breed's Hills, on June 17, 1775. American militias occupied Breed's Hill and fired down on the British as they marched up the hill. On the third attempt, the British finally made it to the top. But almost half of the 2,400 British troops were wounded or killed. That was twice the number of Americans injured or killed.

Though the British won the Battle of Bunker Hill, the battle showed that the Americans were serious about fighting for independence.

In September, the Continental Army under General Phillip Schuyler decided to invade British-controlled Canada. They hoped to get support from the French Canadians who lived there. Ethan Allen tried to capture Montreal with his Green Mountain Boys. But with only 100 men, they were quickly defeated. General Richard Montgomery of the Continental Army eventually captured Montreal on November 13.

Montgomery then moved toward the western side of Quebec. Meanwhile, General Benedict Arnold moved his troops to the eastern side. They planned to meet in the city. But Montgomery's forces were spotted by the British. Montgomery was killed, and some of his troops fled. Arnold's troops came under fire. Still, they breached the city walls. However, when Montgomery's troops did not meet them, Arnold's forces retreated.

The French Canadians had sided with the British. Making things worse, the Continental armies were freezing and faced an outbreak of smallpox. After the Battle of Quebec, the Americans retreated from Canada.

Benedict Arnold would later betray the Continental Army, joining the British.

The Battle of Trenton was a relatively small battle, but it was important in the course of the overall war.

By late 1776, the morale of the Continental Army was low. The British had pushed Washington's troops back to Pennsylvania. The Continentals were camped in freezing weather on the banks of the Delaware River. Washington came up with a daring plan. On December 25, he moved his troops across the river in a storm and marched them several miles to Trenton, New Jersey. Here, they surprised and defeated 1,500 Hessians, the German troops hired to fight for Britain. Washington's forces then moved to capture Princeton, New Jersey. These successes gave the Americans the hope they needed to go on.

# THE TURNING POINT

The Battle of Saratoga in New York began on September 19, 1777. The British hoped to control New York's Hudson Valley, an important gateway into the country. After successfully taking Fort Ticonderoga, the British marched toward Saratoga. Here, American forces had built defenses overlooking the Hudson River. After a fierce battle, the British surrendered to General Horatio Gates on October 17. This victory became a turning point in the war.

American writer, scientist, and diplomat Benjamin Franklin was in France seeking help for the revolution. The victory at Saratoga convinced France to sign the Treaty of Alliance with the United States in February 1778. The French were already secretly supplying the Americans with weapons and money.

British general John Burgoyne surrendered his forces at Saratoga, giving the Continental Army a key victory.

Following fierce fighting at the Battle of Yorktown, where US and French forces assaulted British fortifications, the United States emerged victorious in its revolution.

Some help came from the Marquis de Lafayette, a French military officer who arrived in the United States in 1777 to fight for Washington. Now, France would officially provide both land and naval support.

The French fleet helped guard the coastlines, cutting off British supplies. In 1780, the French general Rochambeau arrived in Rhode Island with 5,000 French soldiers. He joined forces with Washington. In August 1781, they moved into Virginia toward British general Charles Cornwallis's army in Yorktown. Lafayette blocked the British escape routes. Washington's army surrounded Yorktown. Cornwallis surrendered on October 19, 1781. The United States had won its independence.

# A NEW COUNTRY

The Treaty of Paris, signed on September 3, 1783, formally ended the war. Britain recognized the independence of the United States as far west as the Mississippi River. The treaty maintained British control of Canada.

The Continental Congress had created a central government under the Articles of Confederation in 1777. But problems arose from the federal government's lack of power. In 1787, leaders developed a new government at the Constitutional Convention. The result was a new framework in

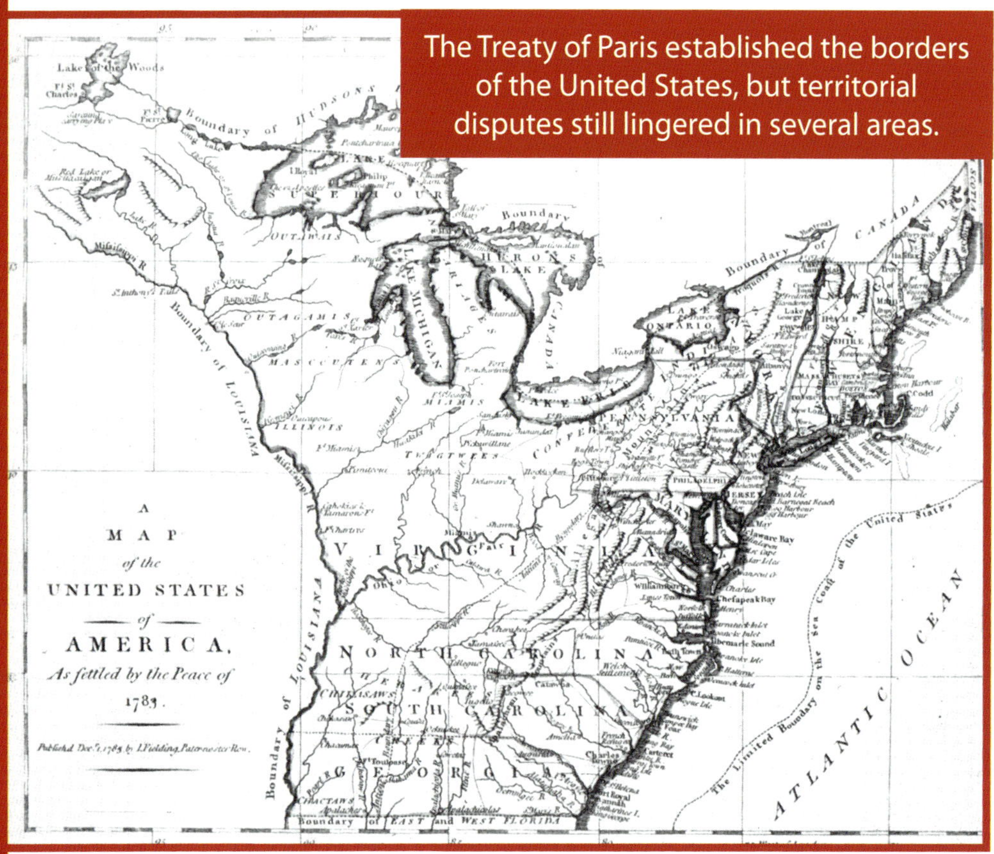


The Treaty of Paris established the borders of the United States, but territorial disputes still lingered in several areas.

The National Archives in Washington, DC, houses the original copies of the US Constitution and the Declaration of Independence.

the US Constitution. This document gave the federal government the right to raise money, keep an army, and print money. Many other powers remained with the states. The Bill of Rights added ten amendments to the Constitution spelling out the rights of the people.

## AT A GLANCE

**Colonial Forces**: 231,000 Continental Army troops; 145,000 militia troops
**Colonial Military Deaths**: 6,800 in battle; 17,000 from disease
**Civilian Deaths**: unknown

# WAR OF 1812 (1812–1815)

The War of 1812 was fought between the United States and Great Britain, now known as the United Kingdom. In Europe, the United Kingdom was at war with the French emperor Napoleon Bonaparte. The British blockaded French ports so their enemies could not get supplies. The United Kingdom suspected that the United States was helping France, so the British boarded US ships to check for military supplies. The British also forced US sailors to serve on British ships, a

Following the American Revolution, the United States built up its military strength, constructing ships to defend the young nation.

The victory of the *Constitution*, *left*, against the *Guerriere* has become a celebrated part of US Navy history.

practice called impressment. The United States demanded that this stop. The British ignored these requests. On June 18, 1812, the United States declared war on the United Kingdom.

Despite the United Kingdom's superior forces, the United States had some early victories. The USS *Constitution* under the command of Captain Isaac Hull fought the British ship HMS *Guerriere* in the North Atlantic in August 1812. The *Constitution*'s sturdy oak frame and sides seemed to repel cannonballs. This earned the ship the nickname Old Ironsides. The *Guerriere* surrendered, and the *Constitution* returned to cheering crowds in Boston.

Mounted troops, known as cavalry, played an important role in the US Army of the 1800s.

The United States tried to invade British-controlled Canada in the summer of 1812. Its forces lost the Battle of Queenston Heights on the Niagara River and were unable to take Quebec. Fort Detroit in the Michigan Territory surrendered to the Canadians. The British joined forces with some American Indian groups, providing them with weapons to

fight American settlers on the frontier. The fighting along the Canadian border and around the Great Lakes was bloody. US forces raided and burned towns. In revenge, British soldiers bayoneted surrendering US soldiers at Fort Niagara.

In 1813, the United States had two important victories. In September, a fleet of nine US ships defeated six British warships on Lake Erie. This forced the British to abandon Detroit and gave Americans naval control of the Great Lakes. In October, US forces defeated American Indian soldiers under the Shawnee leader Tecumseh at the Battle of the Thames near Detroit.

Though he lost his own ship early in the clash, US naval commander Oliver Hazard Perry was named a hero for his victory over British ships on Lake Erie.

## ATTACK ON THE CAPITAL

A temporary peace in Europe allowed the British to bring more forces to fight the Americans. In August 1814, General Robert Ross led 4,500 British troops into Maryland. He quickly moved to capture Washington, DC. The army burned government buildings, including the White House and the US Capitol.

The British land forces then moved to capture Baltimore, Maryland. The Maryland militia successfully defended the city,

When the British burned Washington, DC, President James Madison was forced to flee the capital.

Francis Scott Key, *top*, saw the battle at Fort McHenry and was inspired to write a poem about it. The poem was later set to music and became the US national anthem.

and General Ross was killed. American forces also held off the British navy at Fort McHenry. The battle lasted 25 hours and inspired the writing of the US national anthem, "The Star-Spangled Banner." The British gave up on Baltimore and attacked New Orleans, Louisiana.

## THE US VICTORY

As fighting began in New Orleans, representatives from the United Kingdom and the United States signed the Treaty of Ghent to end the war in December 1814. The United Kingdom was willing to end the war so that it could focus on conflict closer to home in Europe. However, news traveled slowly at the time, and word of the treaty took weeks to reach New Orleans. US general Andrew Jackson defeated British forces outside the city on January 8, 1815. The war formally ended on February 18 when President James Madison signed the treaty.

Andrew Jackson, *in blue, holding sword*, became famous for his victory at the Battle of New Orleans and was later elected president.

The Uncle Sam figure, made famous in this World War I recruiting poster, had his origins in the War of 1812.

The war changed little. The Treaty of Ghent returned all territory of the United States and United Kingdom to its pre-war status. However, the War of 1812 brought respect to the United States for its ability to defend itself. People felt patriotic as the different states came together to defeat a common enemy. New American symbols, such as the patriotic figure known as Uncle Sam, emerged from the war. The war also set the stage for future expansion.

## AT A GLANCE

**US Forces**: 35,000 US forces; 458,000 state militia
**US Military Deaths**: 15,000
**Civilian Deaths**: unknown

# AMERICAN INDIAN WARS (1790–1891)

Before the United States became a country, European settlers had taken many American Indian territories along the eastern coastline. When the United States formed, its leaders signed treaties with many remaining tribes to protect tribal lands. But as more settlers demanded land, the federal government did not keep its promises to the tribes. On May 28, 1830, President Andrew Jackson signed the Indian Removal Act. This gave the president the right to grant prairie lands west of the Mississippi River, which US settlers did not want, to American Indians who gave up their homelands in the east.

The tribes did not want to trade their good land and traditional homes for unknown land in this region, which became known as Indian Territory. Still, the US Army forced approximately 100,000 men, women, and children west. The Choctaw went first, followed by the Chickasaw and Creek. When the government tried to remove the Seminole from Florida, a war began that lasted seven years. But the Seminole eventually lost.

The last people forced out were the Cherokee, who lived in parts of Tennessee, Georgia, Alabama, and North Carolina. In 1835, a small number of Cherokee signed the Treaty of New

A 1942 painting by Robert Lindneaux depicts the devastating Trail of Tears.

Echota, which promised the Cherokee new lands and $5 million for their old lands. In 1838 and 1839, the US Army forced the Cherokee more than 1,000 miles (1,600 km) west to Indian Territory, in what is now the state of Oklahoma. The journey, known as the Trail of Tears, began in Tennessee in late fall. The people traveled, mostly on foot, all winter. Many lacked the clothing and equipment needed for the difficult journey. They arrived in March. Historians estimate that 20 percent or more of the travelers died along the way.

Fort Laramie was located in what later became southeast Wyoming.

## THE PLAINS WARS

The Great Plains region was home to more than 30 American Indian tribes. In 1851, the United States signed the Fort Laramie Treaty with tribes to protect American Indian lands from US settlers traveling west. Still, settlers who were headed to the gold fields of California often intruded on American Indian lands.

The northern Great Plains was the homeland of the Sioux. In 1854, settlers claimed someone from the Lakota tribe, a subgroup of the Sioux, had killed a cow belonging to the settlers. When the Lakota refused to turn over the suspect to the US Army, soldiers opened fire. The Lakota retaliated by killing the soldiers. The conflict continued, and the next year, the US Army defeated the Lakota at Blue Water Creek in present-day Nebraska.

Fighting continued up and down the Plains, with US forces striking at Comanche settlements. In May 1858, Captain John S. Ford and a frontier military force called the Texas Rangers attacked Comanche chief Iron Jacket's camp. In October, the Army planned to attack Chief Buffalo Hump's camp. But instead, the soldiers accidentally attacked a village of Comanche people who were seeking peace.

In 1862, tensions rose between the Dakota in Minnesota and white settlers. The Dakota people there struggled with failing crops, disappointing hunting, and delays in payments they were owed by the US government. Beginning that August, a group of Dakota led by Little Crow killed hundreds of white settlers over a period of several weeks. In September of that year, the Army defeated the Dakota at the Battle of Wood Lake. Little Crow and many of his warriors escaped and fled west. The Army went after them, killing many warriors and destroying their supplies.

An early photograph shows Little Crow in 1862.

ATTENTION!

INDIAN

FIGHTERS

Having been authorized by the Governor to raise a Company of 100 day

U. S. VOL CAVALRY!

For immediate service against hostile Indians. I call upon all who wish to engage in such service to call at my office and enroll their names immediately.

Pay and Rations the same

Volunteer Caval

Parties furnishing their own horses will receive 40c per d

while in the service.

The Company will also be entitled to all horses and other p

Office first door East of Bec

Central City, Aug. 13, '64.

The government posted notices like this to recruit people to fight against American Indians in the West.

## MORE BROKEN TREATIES

The Cheyenne chief Black Kettle wanted peace. Many times, he had negotiated with the government. After violence increased, he tried again in 1864 to end the war. The Army told him to return to Sand Creek, in what is now Colorado, to await a decision. Here, Army troops attacked and slaughtered 150 to 200 Cheyenne and Arapaho people.

Tribes had few supplies and warriors, and the United States was tired of fighting amid the American Civil War. In 1865, the United States signed new peace treaties with several American Indian tribes. However, the treaties did not last long. The American Indian leaders who signed them did not have authority over the many Plains nations, and the US government was unable to stop the flood of settlers from moving west.

An Oglala chief, Red Cloud, gathered a coalition of tribes to defend their territories along the Bozeman Trail, which ran between what are now Wyoming and Montana. They killed 80 men from Fort Phil Kearny near the Bighorn Mountains in December 1866. In the Treaty of Fort Laramie, signed in 1868, the Army agreed to leave its forts along the Bozeman Trail.

The December 1866 clash pitted Lakota, Cheyenne, and Arapaho warriors against US Army troops under Captain William Fetterman.

Farther south, General Philip H. Sheridan laid a trap for the American Indian groups that resisted living on reservations, areas set aside where the US government forced many American Indians to live. Fighting in what is now Oklahoma, Lieutenant Colonel George Custer killed Black Kettle and his followers. Many surviving American Indians from the Kiowa, Comanche, Arapaho, and southern Cheyenne tribes agreed to move onto reservations.

By 1874, the United States began surveying land for the Northern Pacific Railroad. The Army was sent to protect

An engraving depicts Custer's attack on the camp of Black Kettle.

In what became known as Custer's Last Stand, Lakota, Cheyenne, and Arapaho warriors overwhelmed Custer's troops at Little Bighorn.

the area. The Lakota, led by Crazy Horse, and the Cheyenne, led by Sitting Bull, formed an alliance. They had early victories. In June 1876, Custer and nearly half his men died at the Battle of the Little Bighorn in southern Montana. The Army retaliated, forcing the tribes west. Sitting Bull fled to Canada for a time.

## APACHE WARS

In 1861, a raiding party from the Arivaipa band of Apache attacked a settler's farm in Arizona, stealing livestock and kidnapping a child. In response, the Army arrested Cochise, the leader of a different band of Apache, who was not involved in the incident. He escaped, but US forces took several of

A famous 1889 painting by Frederic Remington depicts white horsemen fleeing from Apache warriors during the Apache Wars.

his family members hostage, eventually killing them in the following days. This led to the Apache Wars.

These conflicts featured few major battles. Instead, the two sides fought in smaller raids and ambushes. Cochise finally surrendered in 1872, agreeing to move his people to the Chiricahua Reservation in southeast Arizona.

# THE GHOST DANCE MOVEMENT AND WOUNDED KNEE

The Lakota were promised 60 million acres (24.3 million ha) of land in what is now South Dakota in 1868. By 1887, the General Allotment Act reduced the Lakota land to only 12.7 million acres (5.1 million ha). Some Lakota formed the Ghost Dance religious movement, which said the lands would soon return to American Indian peoples. Feeling threatened by this movement, the US agent for the reservation sent police to arrest Sitting Bull, who now lived

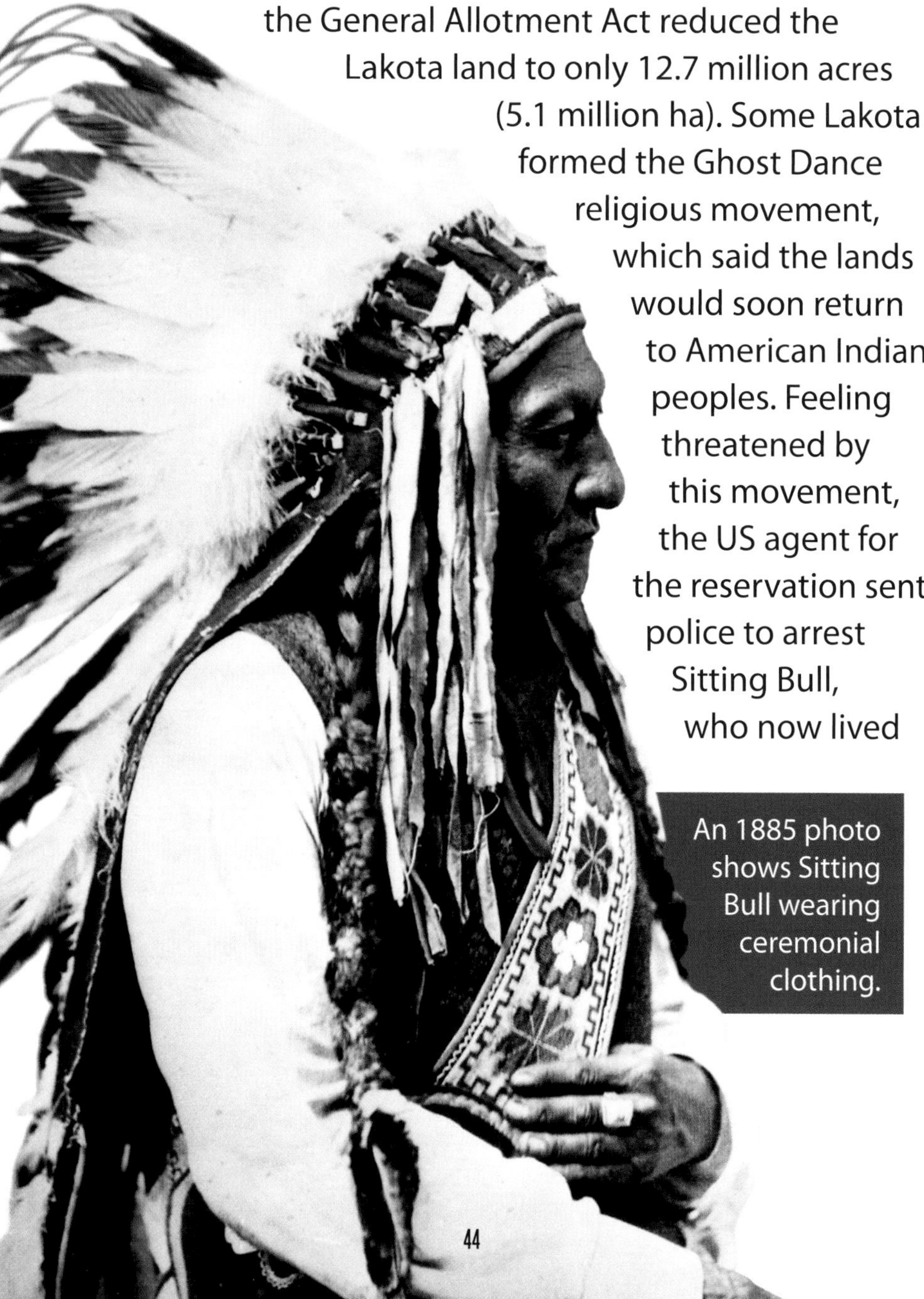

An 1885 photo shows Sitting Bull wearing ceremonial clothing.

Today a monument marks the location of the Massacre at Wounded Knee.

on the reservation. Sitting Bull was not a Ghost Dancer, but some of his followers were. Violence broke out. Several Lakota, including Sitting Bull, were killed.

On December 29, 1890, the Lakota chief Big Foot was leading a group to the Pine Ridge reservation in search of peace. Suddenly, they were surrounded by cavalry at Wounded Knee Creek in southwestern South Dakota. As the soldiers tried to take the Lakotas' weapons, a gun accidentally went off. The cavalry opened fire. Soldiers killed between 250 and 300 Lakota, half of them women and children. The Massacre at Wounded Knee was the last major violent incident of the American Indian Wars.

## AT A GLANCE

**US Forces**: 106,000
**US Military Deaths**: 1,000
**US Civilian Deaths**: 19,000
**American Indian Deaths**: 30,000

# MEXICAN-AMERICAN WAR (1846–1848)

The Mexican-American War began over a border dispute. In 1836, Texas split from Mexico to become an independent nation. The United States then annexed Texas in 1845. However, Mexico claimed Texas ended at the Nueces River. The United States claimed it ended 100 miles (160 km) farther to the

A disagreement about the status of the Rio Grande River as the border between the United States and Mexico helped lead to the Mexican-American War.

President James K. Polk campaigned for president on a promise of expanding the nation's territory.

southwest, at the Rio Grande River. Mexico cut relations with the United States because of the dispute.

President James K. Polk believed in a doctrine known as Manifest Destiny. This idea held that the United States was meant to occupy North America from the Atlantic to the Pacific. Polk sent John Slidell, the US minister to Mexico, on a secret mission to make a deal. He wanted to negotiate the Texas boundary and offer $30 million to purchase the territory that would later become the states of New Mexico and California. Mexican president José Joaquín Herrera refused to meet with him.

The Battle of Resaca de la Palma ended in an American victory.

Polk was insulted. He ordered General Zachary Taylor to occupy the disputed area between the rivers. Mexican troops crossed the Rio Grande and attacked US troops in the Battle of Resaca de la Palma. Several US troops were killed. Polk asked Congress to declare war since Mexico had invaded what the United States considered its own territory. Congress declared war on May 13, 1846.

People in the United States were divided over the conflict. Many members of Polk's party, the Democrats, favored the war. The Whig Party criticized the war as an excuse to grab land from Mexico. The abolitionists, who wanted to ban slavery, feared that the added land would become states where slavery was allowed.

A young Abraham Lincoln, photographed here in 1846, was opposed to the Mexican-American War.

# INVADING MEXICO

Colonel Stephen Kearny moved his forces into New Mexico and California. He met little resistance. After a few skirmishes south of what would become Los Angeles, the Mexican forces in California signed the Treaty of Cahuenga in January 1847. They agreed to stop resistance, lay down their arms, and accept the laws of the United States. They could then return to their homes in California.

Meanwhile, the US Army under Taylor marched south and took the Mexican city of Monterrey. Taylor, with about 5,000 troops, was greatly outnumbered by Mexican general Antonio

About 120 Americans died in the fighting for Monterrey.

The arrival of US troops in Mexico City signaled the end of the Mexican-American War.

López de Santa Anna, who had 14,000 troops. However, the Mexican troops were poorly armed and trained. In February 1847, the US Army defeated the Mexican forces at the Battle of Buena Vista near Monterrey. As the battered Mexican army fled, Taylor did not go farther south.

Polk was angry that Taylor had not pushed toward Mexico City. He ordered General Winfield Scott to attack the Mexican port of Veracruz on the Gulf of Mexico. The United States took the city and started marching west toward Mexico's capital. US forces met some resistance from Santa Anna's troops. Still, they entered Mexico City on September 14, 1847, ending the fighting.

# THE TREATY OF GUADALUPE HIDALGO

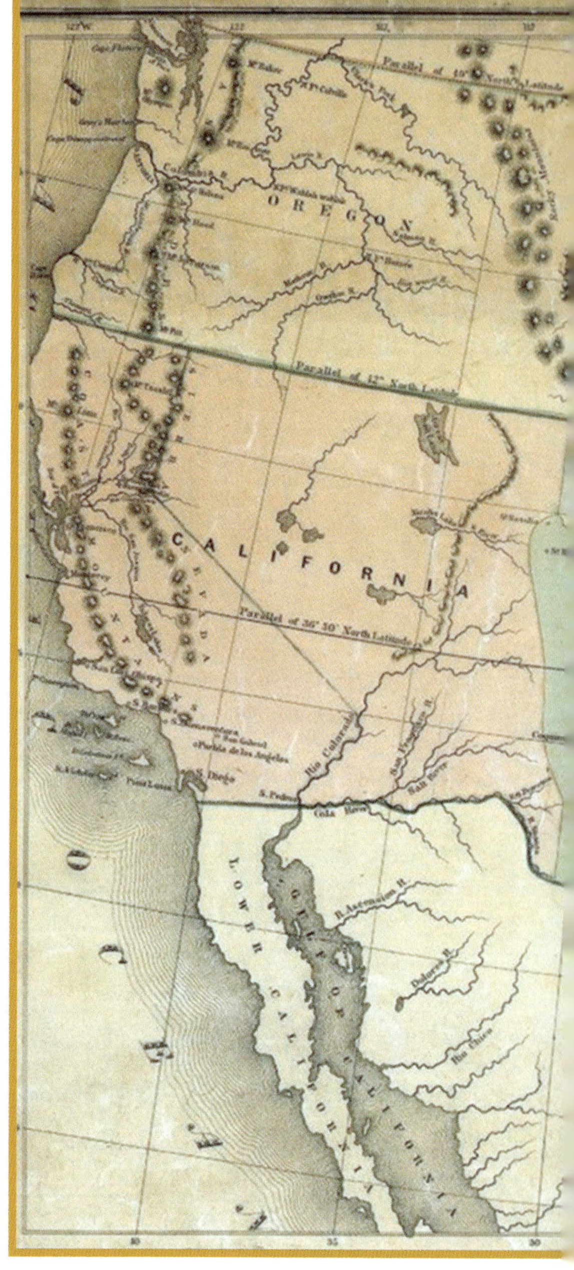


On February 2, 1848, Mexico and the United States signed the Treaty of Guadalupe Hidalgo. Mexico ceded more than 525,000 square miles (1.4 million sq km) of land. This included what is now Arizona, California, western Colorado, Nevada, New Mexico, and Utah. Mexico also agreed to recognize the Rio Grande as the southern border of Texas. Mexican people who lived in the ceded territory kept their land and became US citizens. The United States agreed to pay $15 million to Mexico for the land. The new land drew more US settlers west, attracted by the rich lands and gold mines of California.

Many people who later became important figures in US history fought in the Mexican-American War. Ulysses S. Grant, who eventually led Union forces in the American Civil War and became president, gained experience in military command.

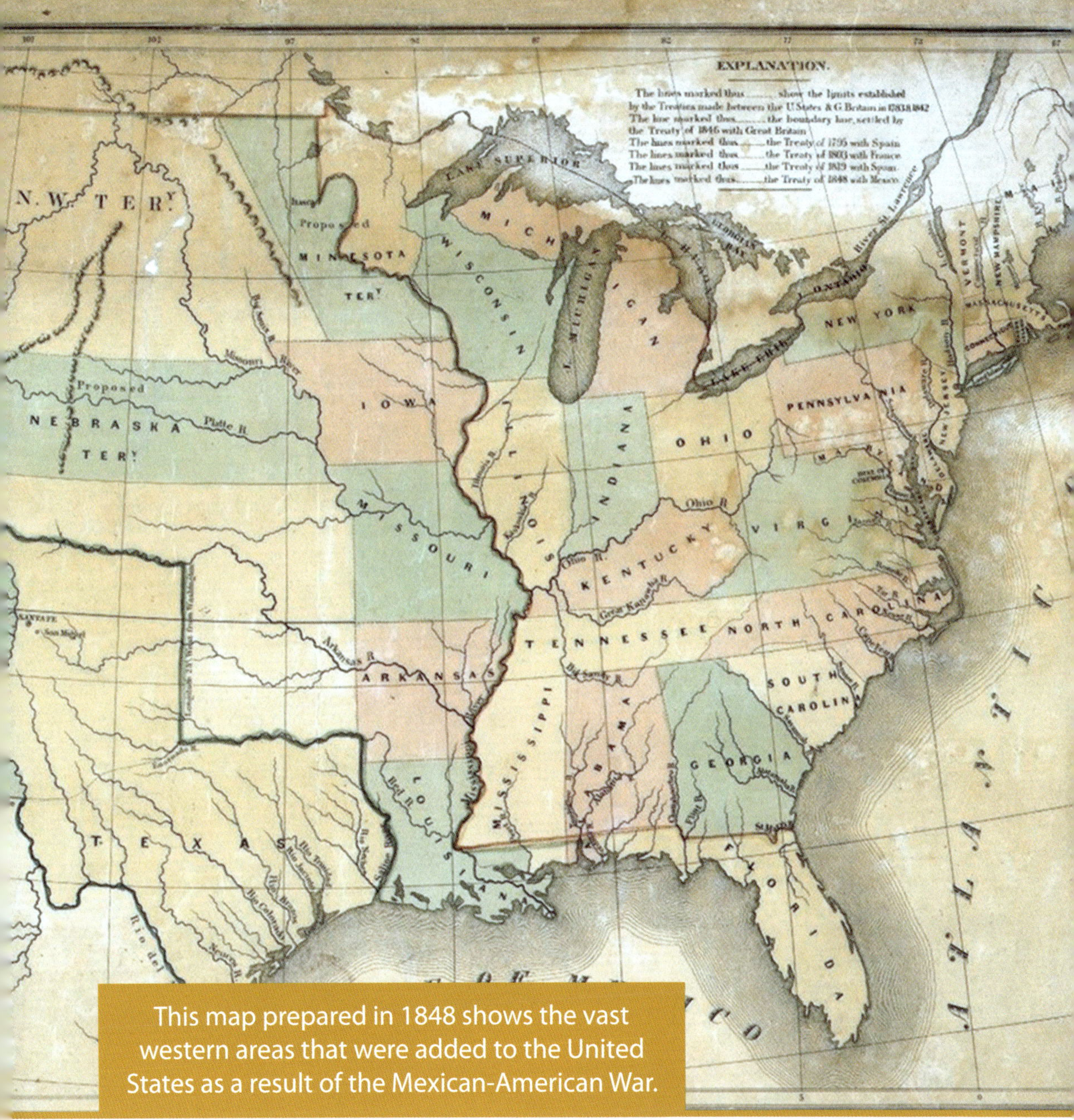


This map prepared in 1848 shows the vast western areas that were added to the United States as a result of the Mexican-American War.

Robert E. Lee, who would become a Confederate general during the Civil War, also fought in the conflict.

## AT A GLANCE

**US Forces**: 78,718
**US Military Deaths**: 13,283
**Civilian Deaths**: unknown

# AMERICAN CIVIL WAR (1861–1865)

The American Civil War was a conflict to decide whether the United States would stay one country or be divided into two countries. The Northern states, known as the Union, wanted to keep the country together. The Southern states wanted to split off to form the Confederate States of America.

At the time of the Civil War, Northern cities such as New York were economic and industrial powerhouses.

The agricultural economy of the South at the time of the Civil War used the forced labor of millions of enslaved Black men, women, and children.

The two regions had very different economies. The North had large cities filled with factories and businesses. The South's economy depended on agriculture, with millions of enslaved Black people forced to work on farms producing crops such as cotton and tobacco.

Slavery was the major issue dividing the states. In 1808, the Act Prohibiting the Importation of Slaves took effect. This made the slave trade with other nations illegal. But slavery continued for enslaved people already in the country. Because the children of enslaved people were automatically enslaved, the total number of enslaved people grew during the following decades. Abolitionists pushed to end all slavery. But Southern states feared that the federal government outlawing slavery would devastate their economies.

Frederick Douglass, who was born into slavery and later escaped to freedom, became a major voice in the abolitionist movement.

Lincoln's inauguration was held in March 1861 in front of the US Capitol, whose dome was still under construction.

Abraham Lincoln, a member of the antislavery Republican Party, was elected president in 1860. Four days after the election, South Carolina senator James Chesnut resigned and returned to his home state. Others followed, and soon 25 of the US Senate's 66 members were gone. Seven states seceded from the United States: South Carolina, Mississippi, Florida, Alabama, Georgia, Louisiana, and Texas. Jefferson Davis, a former US senator from Mississippi, became the first president of the Confederacy.

The Battle of Bull Run, the first major clash of the Civil War, was a harsh defeat for the Union forces.

## THE FIRST SHOTS

Confederate troops fired on Fort Sumter in South Carolina on April 12, 1861. Union soldiers surrendered the next day. No one was killed. However, the Confederacy had won the first battle of the war. President Lincoln put out a call for 75,000 militia members to retake US property and help preserve the Union.

In July 1861, Lincoln pushed General Irvin McDowell of the Union army to capture the Confederate capital in Richmond, Virginia. McDowell's troops set out from Washington, DC.

They were stopped by General P. G. T. Beauregard's Confederate forces at Bull Run, a creek approximately 25 miles (40 km) from Washington, DC. The Union forces retreated, at times obstructed by large numbers of civilians who had come to watch the battle.

General Robert E. Lee became commander of the Confederate army in 1862. His forces won some early battles in Virginia. This encouraged the Confederacy to continue fighting against the superior numbers of the Union. Lee wanted to move the fighting into Union territory and take Washington, DC. He hoped this would encourage European nations to support the Confederacy. Lee invaded Maryland in September 1862.

Robert E. Lee, a veteran of the Mexican-American War, became a leading general for the Confederacy.

In the fighting at Antietam, Maryland, Union troops under Major General George B. McClellan outnumbered the Confederate forces almost two to one. It became the deadliest one-day battle of the war. More than 21,000 soldiers were killed or wounded. The outcome was a draw, and Lee's forces escaped to the South.

The bloody Battle of Antietam involved more than 130,000 troops in all.

The Union navy successfully shut down Confederate ports, a key part of the overall Union strategy.

On January 1, 1863, Lincoln issued the Emancipation Proclamation. The proclamation declared that all slaves in the Confederate states were free. The Emancipation Proclamation also allowed Black soldiers to fight for the Union military. Approximately 180,000 Black Americans joined the Union army and 20,000 joined the Union navy.

At sea, the Union sought to cut off the Confederacy's supply lines. The Union navy blockaded 3,500 miles (5,600 km) of Confederate coastline. Ships stopped the Confederacy from exporting goods, such as cotton, to earn money. The blockade also prevented military supplies from coming in.

## GETTYSBURG

Lee next marched his troops north to Pennsylvania, moving around Washington, DC. His objective was to win a battle in the North, hoping this would lead to negotiations to end the fighting. Lincoln ordered General George G. Meade to prevent Lee's army from entering Washington, DC. Meade followed Lee north.

General George G. Meade led Union forces in several of the war's important battles.

The three-day clash at Gettysburg was the deadliest battle of the war.

On July 1, 1863, a Confederate division marched toward Gettysburg, Pennsylvania, for supplies. They were met by Union forces. The 30,000 Confederates pushed the 20,000 Union soldiers south of Gettysburg. The next day, the Confederates tried to surround the Union troops. Through fierce fighting, the Confederates forced the Union soldiers onto Cemetery Ridge. But the Union held until dark. On the third day, Lee ordered General George E. Pickett to take the Union position on Cemetery Ridge. During what became known as Pickett's Charge, the Union soldiers destroyed the attacking Confederate troops. Lee moved his remaining forces back to Virginia.

Approximately 51,000 soldiers were killed, wounded, missing, or captured at Gettysburg. Meade did not pursue the Confederate army as it fled south. Gettysburg was the turning point of the Civil War in favor of the Union.

## FIGHTING ON OTHER FRONTS

In 1863, Union general Ulysses S. Grant was trying unsuccessfully to take Vicksburg, Mississippi. The capture of the town would allow the Union to control the Mississippi River, which was vital for transport. Vicksburg was well defended. The Union tried attacking with ships on the river, but high bluffs kept the city secure. A land assault failed due to the maze of

swamps protecting the city. In February and March 1863, Grant even tried digging a canal to divert the Mississippi River.

Grant finally succeeded in taking the city by marching around it and assaulting from the southeast. On May 18, the Union army lay siege to the 30,000 troops isolated in Vicksburg. Running out of food and ammunition, the Confederates surrendered on July 4.

Following unsuccessful direct attacks, Grant's troops dug in for a lengthy siege of Vicksburg, Mississippi.

In Georgia, Union general William T. Sherman's troops captured and burned Atlanta, a vital rail and manufacturing center. Atlanta surrendered on September 2, 1864. Sherman then began a march to the sea to capture the port of Savannah. The Union forces marched quickly through the Georgia farmland. Few civilians were harmed, but the soldiers destroyed Confederate buildings, railways, bridges, and more on the way.

Sherman's plan had two purposes. First, destroying infrastructure would make it harder for the Confederate forces to regroup, resupply, and fight back. Second, the destructive

Destroying infrastructure, such as railroad tracks, was an important part of Sherman's strategy.

As word spread of Lee's surrender, the remaining Confederate forces in the field also laid down their arms.

march would strike fear into the people of the Confederacy. By showing overwhelming force, Sherman hoped to destroy the South's will to fight. Sherman's army traveled 285 miles (459 km) in 37 days.

In early 1865, Union troops gradually surrounded Lee's forces. Lee finally surrendered in Appomattox Court House, Virginia, on April 9, 1865. The war's major fighting was over. A Confederate supporter, actor John Wilkes Booth, shot President Lincoln on April 14. Lincoln died the next day, and Vice President Andrew Johnson became president.

The Battle of Hampton Roads saw history's first combat between ironclad warships, with the Union's *Monitor*, *right*, fighting the Confederacy's *Merrimack*.

# DEATH AND DESTRUCTION

The Civil War was the deadliest war ever fought by the United States. Technological advances, including more accurate rifled muskets, resulted in many casualties. The war also saw developments in armored warships, and the Confederates achieved the first sinking of a ship using a submarine.

At the beginning of the war, the armies were not prepared for the number of wounded. Injured men sometimes lay on the battlefield for days. Many wounded soldiers required amputations. Of the 30,000 amputations performed on Union soldiers, 26 percent resulted in death. Antibiotics had not yet been discovered. Poor sanitation, poor diet, and little understanding of infection led to many illnesses and deaths. Gradually, military medicine improved. The government built hospitals in large cities.

Deaths were another problem. There were no identifying tags for soldiers. Bodies piled up after battles, spreading disease. Volunteers tried to identify dead soldiers and notify their families. Military nurse Clara Barton established an office to locate missing men at the end of the war. In three years, she provided information about 22,000 soldiers.

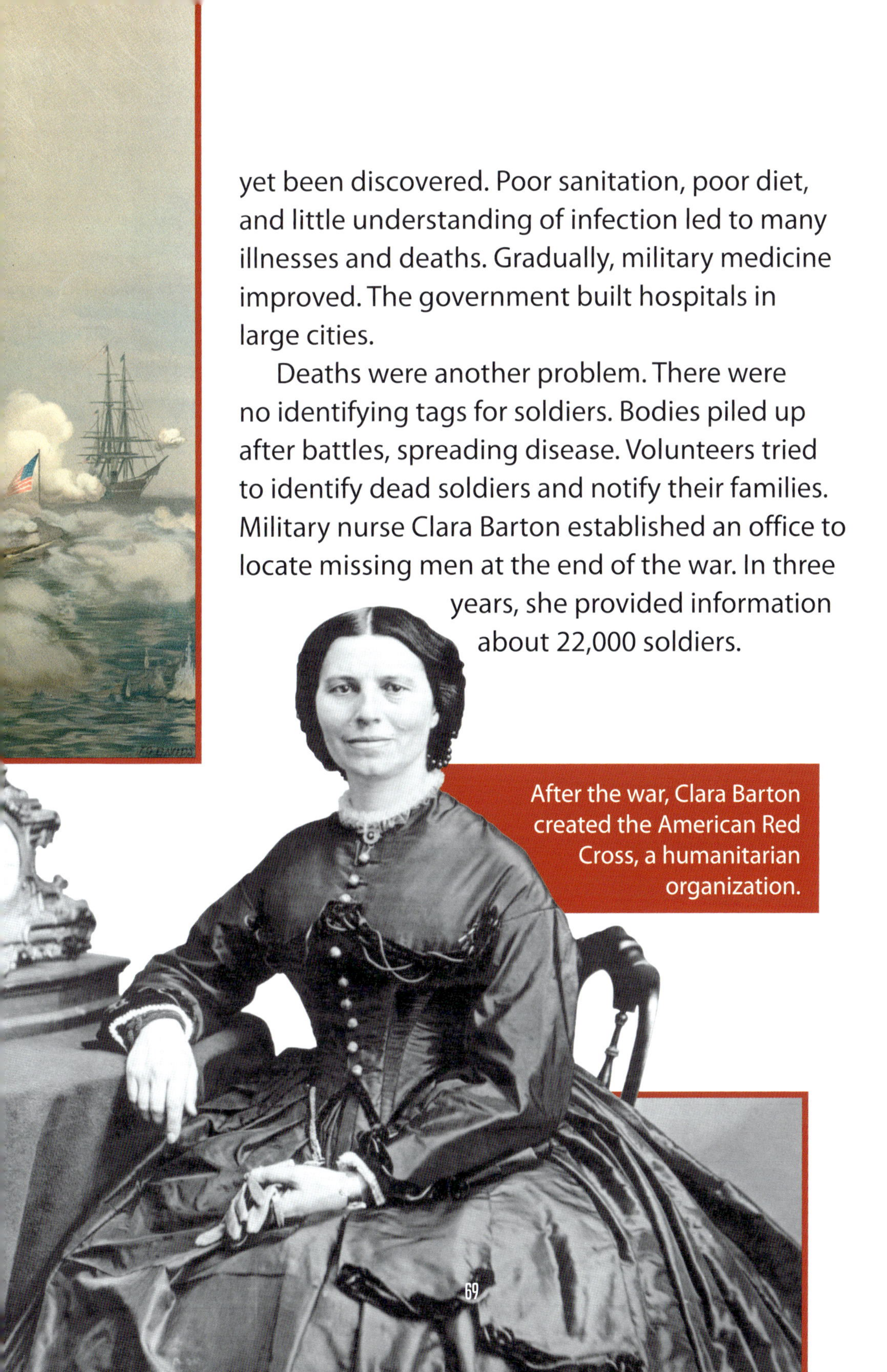

After the war, Clara Barton created the American Red Cross, a humanitarian organization.

Thousands of Union soldiers are buried at Gettysburg National Cemetery.

In 1862, the US Congress gave the president the power to purchase land for national cemeteries. Several of these were created near the sites of large battles, such as Gettysburg. The Union sent units of soldiers into the South to locate graves of Union soldiers and have them reburied in the North. They located 303,536 bodies for reburial, 30,000 of them Black soldiers. The federal reburial efforts were only for Union soldiers. Confederate civilians had to find and rebury their own dead.

Besides advances in weapons and medicine, the Civil War saw a number of other innovations. It was the first major war documented with photographs. Photographer Mathew Brady had a huge impact on society with his images of battlefields during the war. His studio displayed the bloody aftermath of Antietam. These were the first pictures of bodies on a battlefield shown to the public.

Photographs of the gruesome fighting at the Battle of Antietam helped bring home the reality of war like never before.

## CONSEQUENCES

The period from 1865 to 1877 is called Reconstruction. It began with the end of the Civil War and ended when the last US military administrators left the South. Reconstruction was a time to restore the union, rebuild the South, and secure the rights of Black Americans. Public education for both Black and white children was introduced in the South. But this period also brought about white supremacist organizations such as the Ku Klux Klan.

Three Constitutional amendments were passed after the Civil War to protect Black citizens. In 1865, the Thirteenth Amendment abolished slavery in the United States. In 1868,

Much of the South's infrastructure lay in ruins after the Civil War.

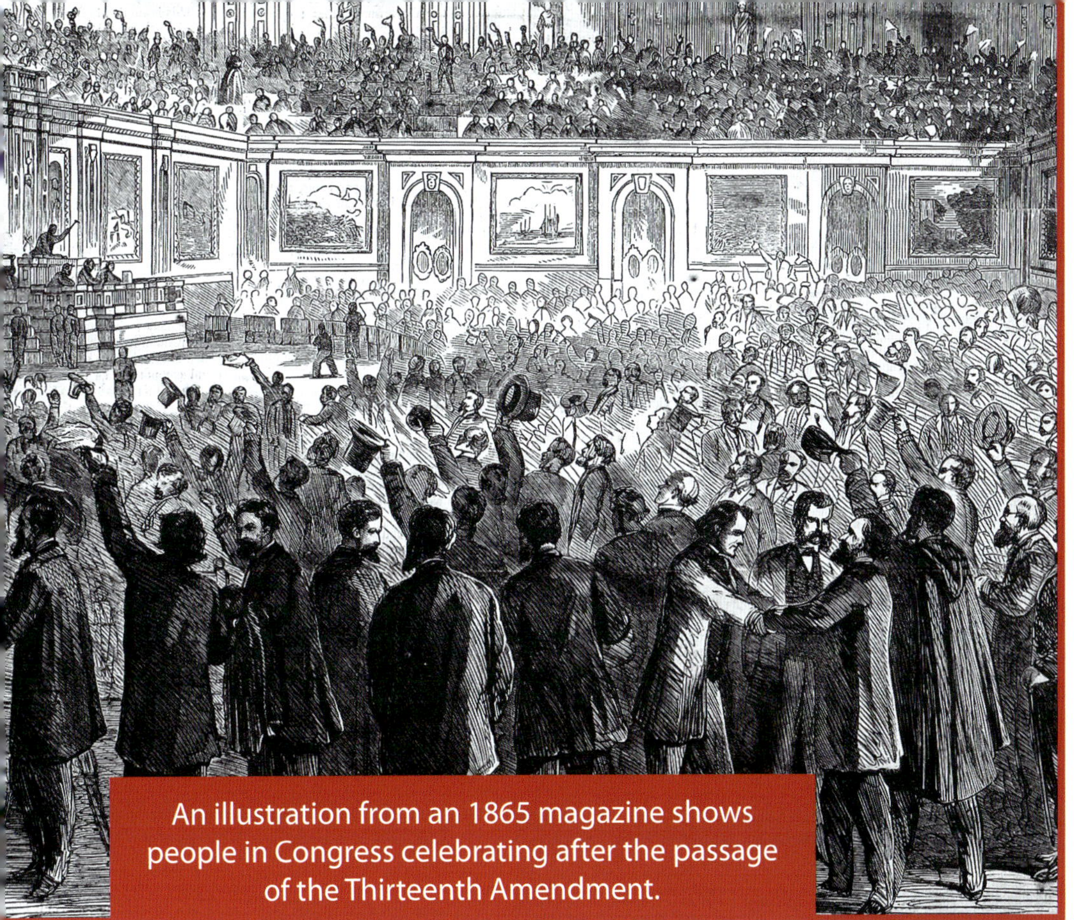

An illustration from an 1865 magazine shows people in Congress celebrating after the passage of the Thirteenth Amendment.

the Fourteenth Amendment provided citizenship to all people born in the United States, including formerly enslaved people. All citizens were guaranteed due process and equal protection under the law. In 1870, the Fifteenth Amendment said that a citizen's right to vote could not be denied due to race, color, or previous condition of servitude. However, many Southern states ignored these protections, creating laws to prevent Black people from voting and fully participating in society.

## AT A GLANCE

**Union Forces**: 1,556,000
**Union Deaths**: 364,511
**Confederate Forces**: 800,000
**Confederate Deaths**: 133,821
**Civilian Deaths**: 50,000

# SPANISH-AMERICAN WAR (1898)

In 1898, Cuba and the Philippines were colonies of Spain. Many people in these colonies wanted independence. American newspapers featured stories of the brutal suppression of the Cuban people by the Spanish. Plus, American owners of sugar businesses in Cuba felt threatened by Spanish control of the island.

Theodore Roosevelt, the assistant secretary of the Navy, was a strong supporter of Cuban independence. When American representatives in Cuba asked President William McKinley for protection for US citizens on the island, Roosevelt sent the USS *Maine*. The ship arrived in Havana, Cuba, in January 1898.

The USS *Maine* had been commissioned in 1895, just a few years before its fateful journey to Cuba.

Three weeks later, two explosions sank the ship. The United States lost 262 sailors. Although no reason for the explosion was found, the United States blamed Spain. US newspapers rallied the public against Spain, and the United States declared war on April 25.

The explosion of the USS *Maine* set into motion the events that led to war.

Roosevelt left his position in the Navy in May 1898 to join a volunteer cavalry unit called the Rough Riders. The unit was a mixture of hunters, cowboys, American Indians, and even college athletes. They joined with the Army Fifth Corps to fight against Spain in Cuba.

Roosevelt and his Rough Riders posed for a photo after the Battle of San Juan Hill.

## WAR IN CUBA

After a blockade of the coasts, US forces arrived in Cuba by June. They moved toward Santiago, Cuba's second-largest city. They quickly reached two Spanish-controlled hills outside the city, called San Juan Hill and Kettle Hill. Several units attacked the two hills in a fight that became known as the Battle of San Juan Hill. Roosevelt led the Rough Riders up Kettle Hill on his horse. The fight also involved Black American cavalry troops

nicknamed the Buffalo Soldiers. The battle ended on July 1 with a US victory.

After fighting ceased in Santiago, Spanish admiral Pascual Cervera y Topete tried to escape Santiago Bay with the Spanish fleet on July 3. The US fleet chased them down and destroyed the Spanish ships. Santiago surrendered on July 17, ending the fighting in Cuba. US troops also landed in the Spanish territory of Puerto Rico on July 25. There was little resistance, and the Americans controlled the island by August 13.

The powerful US Navy was victorious over the weaker Spanish fleet in the Caribbean.

## FIGHTING IN THE PHILIPPINES

The war also extended to the Philippines, a group of islands under Spanish possession in the Pacific. On May 1, 1898, the US Navy under the command of Commodore George Dewey destroyed Spanish ships at Manila Bay. Dewey blockaded the port and waited for US ground forces to arrive. While he waited, Dewey asked for help from Emilio Aguinaldo, a leader of Filipino insurgents fighting the Spanish, to hold Manila, the capital city.

When US forces arrived, they persuaded the Filipinos to step aside and let the Americans take the city. Secretly, Spain had made a deal with the Americans to surrender Manila in an uncontested battle, rather than to the Filipino forces. The Spanish were concerned that the Filipino troops might try to take revenge after defeating them. They surrendered to the Americans on August 14.

The decisive Battle of Manila Bay resulted in minimal casualties for the American side.

The Treaty of Paris was signed on December 10, ending the war. Spain gave up its claim to Cuba, and the island

became independent. Spain ceded Guam, Puerto Rico, and the Philippines to the United States. Aguinaldo and the insurgents expected to gain independence like Cuba. Instead, the Philippines became a US territory.

# RESULTS OF THE WAR

The Kettle Hill charge by Roosevelt and the Rough Riders became legendary. Roosevelt had a reporter nearby to record his heroics. He then wrote a book, *The Rough Riders*, about his adventure. With his flamboyant personality, Roosevelt became a hero and celebrity. He was elected vice president in 1900. In 1901, he became president of the United States following the assassination of McKinley.

Roosevelt was later elected to another full term as president, leaving office in 1909.

The remains of Spanish fortifications can still be seen in Guam today.

The Philippines became fully independent of the United States after World War II. Guam and Puerto Rico remain US territories today. Their people are US citizens and elect their own governors. However, they cannot vote in federal elections since the islands where they live are not states.

## AT A GLANCE

**US Forces**: 306,760
**US Military Deaths**: 2,446
**Civilian Deaths**: unknown

# PHILIPPINE-AMERICAN WAR (1899–1902)

The Philippine-American War was fought between the United States and Filipino revolutionaries. Filipino fighters led by Emilio Aguinaldo had worked with the United States to drive the Spanish out of the Philippines. However, the Treaty of Paris that ended the Spanish-American War in 1898 did not grant independence to the Philippines. Instead, the Philippines became a US territory.

Emilio Aguinaldo, *on horseback*, led Filipino forces against Spain and then against the United States.

US forces had to contend with tropical conditions in the Philippines.

Aguinaldo declared the Philippines independent on June 12, 1898. In January 1899, he called for a constitutional convention and was named president of the Philippine Republic. The United States refused to recognize the new country.

After the Spanish-American War, US major general Elwell S. Otis had only 12,000 combat forces in Manila. Aguinaldo commanded 40,000 men in the countryside. On February 4, 1899, the Battle of Manila began as US forces clashed with Filipino forces around the city. However, the citizens of Manila did not rise up to support the rebels. Aguinaldo's forces were forced to retreat.

The US military shipped tens of thousands of troops to the Philippines to overwhelm the insurgents.

US forces then moved in every direction from Manila. They seized towns and split the Filipino forces into small groups. The Filipinos began to use guerrilla tactics. They had some notable early victories. But by 1901, the number of US troops had increased to 75,000, providing a large advantage in numbers.

## THE CAPTURE OF AGUINALDO

On March 21, 1901, a secret plan led to Aguinaldo's capture. The plot involved US Army colonel Fred Funston. He recruited a force of Filipino Macabebes. The Macabebes were Filipinos from a province that had fought against the Spanish forces.

They also resisted the insurgent forces after Spain left, siding with the Americans.

On March 21, 1901, a group of Macabebes posed as Filipino reinforcements for Aguinaldo. They marched into the insurgent camp leading a group of prisoners. But the prisoners were actually US troops. Once inside the camp, these troops captured Aguinaldo and his men. Attacks by Filipinos continued, but no new leader emerged.

A 1901 illustration depicts the capture of Aguinaldo.

President McKinley was assassinated in September 1901. Vice President Theodore Roosevelt, who had fought in Cuba during the Spanish-American War, became president on September 14. He did not think the Filipinos were capable of self-rule. He wanted to keep the Philippines as a refueling station for the US Navy. Still, he wanted the bloodshed to stop.

Roosevelt proclaimed a ceasefire on July 4, 1902. He pardoned all the remaining Filipinos who had fought

A US photographer took an image of the moment when a group of Filipino insurgents surrendered their weapons.

Taft later became the president of the United States and a Supreme Court justice.

US forces. Aguinaldo pledged allegiance to the United States. He thereafter wore a black bow tie in mourning for his occupied country.

## COLONIAL GOVERNMENT

In March 1900, President McKinley had appointed William Howard Taft as chairman of the Second Philippine Commission. His job was to organize a government for the Philippines. The next year, Taft became the governor of the Philippines and worked for the economic development of the islands. Filipinos elected their first legislature, the Philippine Assembly, in 1907.

In the next decade, Congress continued to reduce US control over the islands. The Philippine Autonomy Act of 1916, also known as the Jones Act, set a path forward for Philippine independence. Filipinos gradually took control of more of the government. Japan occupied the islands during World War II. The United States fought with the

At a ceremony in 1946, officials lowered the US flag and raised the flag of the Republic of the Philippines.

The beautiful natural landscapes and vibrant cities of the Philippines make the country a destination for visitors from around the world, including from the United States.

Filipinos to free them from Japanese control. The Philippines became independent in 1946. Aguinaldo removed his black tie.

Because the Philippines was a US territory, many emigrants left the islands to live in the United States. Some arrived during the colonial period, but many came after World War II to take advantage of opportunities for education and jobs. By the 2020s, more than two million Filipino Americans lived in the United States.

## AT A GLANCE

**US Forces:** 125,000
**US Military Deaths:** 4,300
**Civilian Deaths:** 200,000

# WORLD WAR I (1914–1918)

World War I began in the southeastern European countries known as the Balkans. Countries in the area, including Greece, Serbia, Montenegro, and Bulgaria, had forced the Ottoman Empire out of Europe. In 1912 and 1913, the countries fought among themselves as they tried to divide up the land they had conquered. Serbia especially wanted to take over land where Slavic peoples were living in the southern part of Austria-Hungary.

LA DOMENICA DEL CORRIERE

Si pubblica a Milano ogni Domenica

Supplemento illustrato del "Corriere della Sera"

Uffici del giornale: Via Solferino, N. 28 MILANO

Anno . . . . . . . L. 5 — L. 10 —
Semestre . . . . . » 2.50 - 5 —

Per tutti gli articoli e illustrazioni è riservata la proprietà letteraria e artistica, secondo le leggi e i trattati internazionali.

Centesimi 10 il numero.

Anno XVI. — Num. 27. 5 - 12 Luglio 1914.

L'assassinio a Serajevo dell'arciduca Francesco Ferdinando erede del trono d'Austria, e di sua moglie.
(Disegno di A. Beltrame).

A 1914 illustration from an Italian newspaper depicts the assassination of Archduke Franz Ferdinand.

The Serbian head of military intelligence, Dragutin Dimitrijević, plotted the assassination of Archduke Franz Ferdinand of Austria-Hungary. The archduke was in line to the throne held by Austria-Hungary's Emperor Franz Joseph. On June 28, 1914, Ferdinand and his wife were assassinated in Bosnia, a region under the control

of Austria-Hungary, by Bosnian Serb Gavrilo Princip. Furious, Austria-Hungary soon declared war on Serbia.

Countries in the rest of Europe had many military alliances with each other. Like dominoes falling, many declarations of war were made in August 1914. Russia came to the defense of Serbia. Germany's leader, Kaiser Wilhelm II, then declared war against Russia. The other countries lined up on different sides. In the end, World War I pitted the Central Powers, led by Germany and Austria-Hungary, against the Allied Powers, led by Russia, France, and the United Kingdom. The United States watched from the sidelines at first. It would not join the conflict until 1917.

# A NEW TYPE OF WAR

Wilhelm II carried out the Schlieffen Plan, a strategy for winning a two-front war devised by a German general a decade earlier. Under this plan, he sought to capture France quickly while the Russians were still gathering their forces. His troops marched through Belgium and into France. In September 1914, French and British forces stopped the German troops. The two sides fought in the First Battle of the Marne just north of Paris.

French infantry charge across a field with bayonets attached to their rifles.

The Germans withdrew several miles, then dug in, creating a network of defensive trenches.

Trench warfare was a new kind of war. Technology such as the machine gun and powerful, accurate artillery made it too deadly for armies to fight in flat, open spaces. A machine gun could fire 600 bullets per minute up to 1,000 yards (914 m). Rapid-firing artillery was much deadlier than old guns that had to be reset after each firing.

To protect their forces, armies dug miles of trenches and dugouts that were protected by barbed wire. Soldiers fought, slept, and ate in these ditches. The trenches were miserable, often full of water, mud, and rats. But they provided protection as artillery shells rained down. To advance, armies had to leave their own trenches and face fire from enemy trenches. It was difficult and deadly. Soldiers often stayed in trenches for months and made little progress.

Germany had the advantage of a well-trained army and excellent commanders.

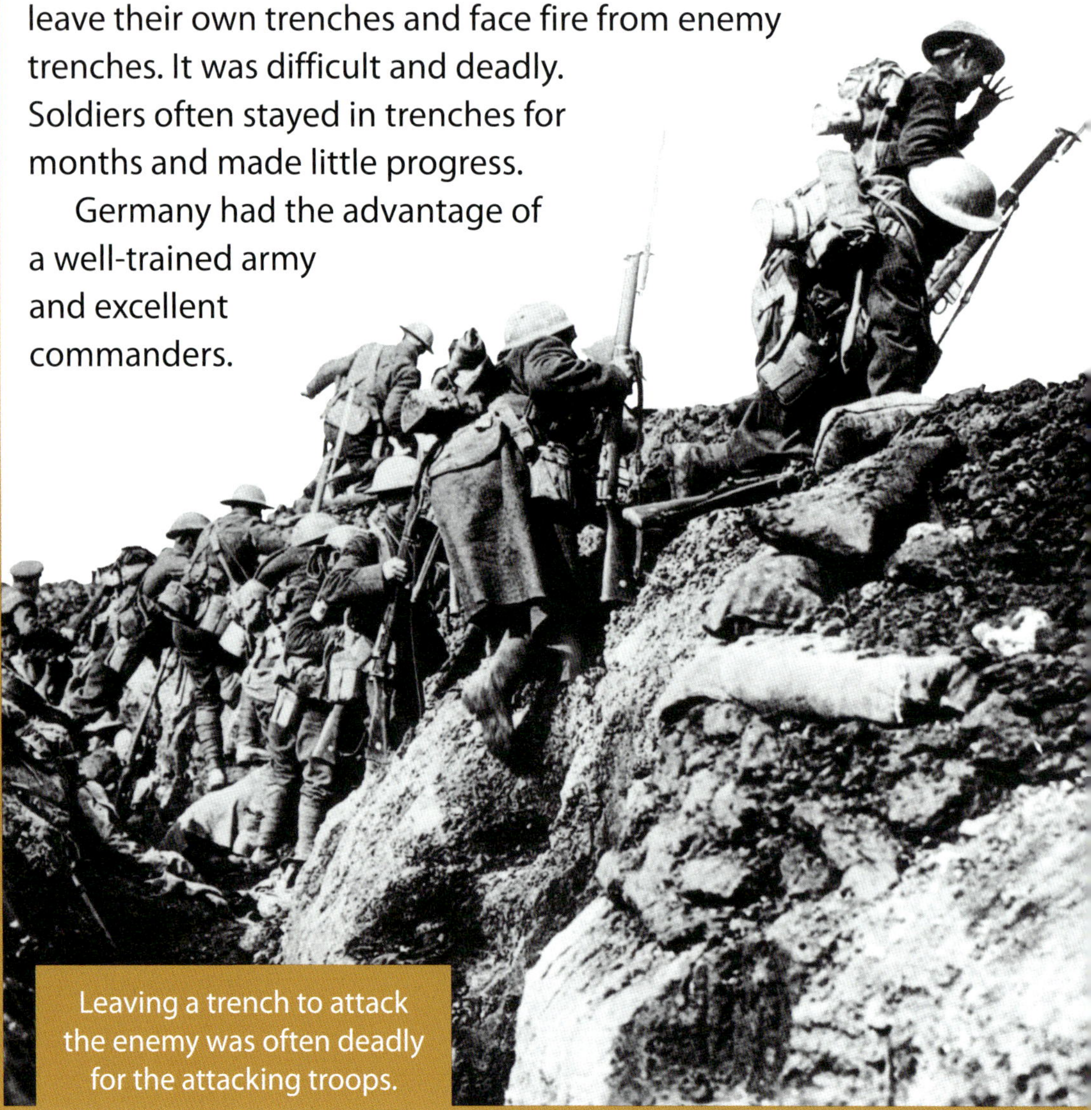

Leaving a trench to attack the enemy was often deadly for the attacking troops.

Gas masks soon became important equipment for soldiers on the front lines.

It possessed good communications and a large railway network. German chemical laboratories also produced a deadly new weapon. On April 22, 1915, chlorine gas released by the Germans poisoned Allied troops. The gas caused choking and could lead to death. The soldiers had no defense against the gas. The success of this attack led both sides of the war to create new poisonous gases. It also led to gas masks and training in gas warfare.

The Allies had some advantages. They had a larger military force. The British navy was the most powerful in the world and successfully kept supplies flowing to the Allies. It could also use blockades to stop supplies from reaching the Germans. But some of these advantages were offset by German technology in submarines and torpedoes.

The Allies landed hundreds of thousands of troops on the Gallipoli Peninsula, but they met fierce resistance and were eventually forced to withdraw.

## BLOODY CAMPAIGNS

One of the first offensives launched by the Allies was the Gallipoli campaign, which began in February 1915. The Allies planned to control the Dardanelles strait and the city of Constantinople in Turkey, key points linking the Mediterranean Sea and the Black Sea. This position would let them control access to the Black Sea and allow them to aid Russia. The Allies began with attacks from the British fleet. They then landed British, Australian, French, and New Zealander troops on the

Gallipoli Peninsula in Turkey. But little progress was made, and the Allies took heavy losses against Turkish troops. Allied military forces withdrew from the area by January 1916.

After being bogged down in the First Battle of the Marne, the next German plan was to crush the French army before reinforcements arrived from the United Kingdom. General Erich von Falkenhayn chose the fortress city of Verdun, France, to attack. He thought destroying the fortress would give the Germans a strategic advantage and demoralize the French.

French troops move toward Verdun on horseback. Horses remained important tools of war through World War I.

On February 21, 1916, the Germans began massive shelling. Although they had early success, both sides were soon fighting from trenches. French general Phillippe Petain made sure his soldiers were well supplied and rotated troops in and out of the trenches. The battle lasted until December 15. Gradually, the Germans withdrew. The French had 377,000 men killed or injured. For the Germans this figure was 330,000 men.

Each side fired vast numbers of artillery shells during the war's major battles.

Photos from observation airplanes showed the networks of trenches and the many craters from artillery strikes that dotted the landscape of the Somme battlefield.

As the French were fighting in Verdun, the British began launching artillery strikes at the Battle of the Somme in June 1916. Their plan was to fire shells until their infantry could attack the German trenches. They fired 1.5 million shells, but the German defenses remained intact. The British infantry attacked on July 1 but were cut down by German machine guns and rifles. The battle continued through the summer.

On September 15, the British tried a new weapon, the tank, at the Somme. This armored vehicle drove on treads instead of wheels, allowing it to clear trenches and rough ground. The British forces advanced, but they didn't break the German lines. On November 18, British general Douglas Haig ordered the end of the attack. After almost five months of fighting, the Allies had only advanced 7 miles (12 km) at a cost of more than one million Allied and German men killed or wounded.

## THE UNITED STATES ENTERS THE WAR

Germany had to cut off supplies to the United Kingdom to win the war. It used submarines to patrol the Atlantic. These submarines sometimes attacked passenger vessels if commanders suspected weapons might be on board. The German subs sank a British passenger ship, the RMS *Lusitania*, in 1915. Nearly 1,200 people drowned,

The first tank to see battle, the British Mark I, looked much different from tanks today.

The sinking of the RMS *Lusitania*, killing many American passengers, drew the United States closer to war.

including 128 Americans. After the sinking of the French ship SS *Sussex* in March 1916, President Woodrow Wilson had enough. He warned Germany not to attack passenger ships and also to allow crews to abandon merchant ships before they were attacked. Germany agreed under a promise called the Sussex pledge.

However, in January 1917, Germany decided to break the Sussex pledge. Wilhelm II was convinced that the submarines could end the war before the United States had time to send troops to Europe. The same month, the British intercepted a telegram from German foreign minister Arthur Zimmermann to the Mexican government. Germany promised to help return the land ceded to the United States after the Mexican-American War if Mexico supported Germany.

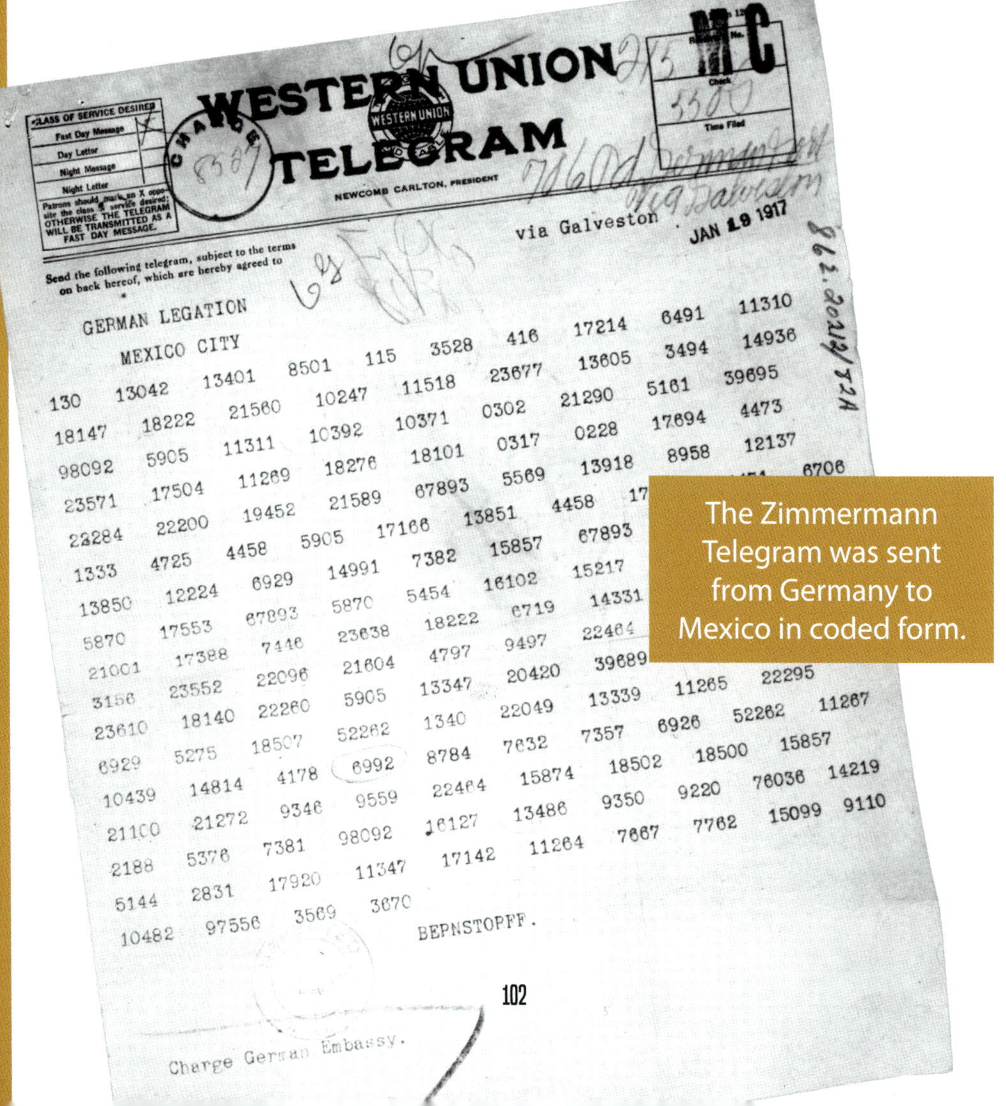

WESTERN UNION TELEGRAM

NEWCOMB CARLTON, PRESIDENT

CLASS OF SERVICE DESIRED
Fast Day Message
Day Letter
Night Message
Night Letter
Patrons should mark an X opposite the class of service desired; OTHERWISE THE TELEGRAM WILL BE TRANSMITTED AS A FAST DAY MESSAGE.

Send the following telegram, subject to the terms on back hereof, which are hereby agreed to

via Galveston JAN 19 1917

GERMAN LEGATION
MEXICO CITY

130 13042 13401 8501 115 3528 416 17214 6491 11310
18147 18222 21560 10247 11518 23677 13605 3494 14936
98092 5905 11311 10392 10371 0302 21290 5161 39695
23571 17504 11269 18276 18101 0317 0228 17694 4473
22284 22200 19452 21589 67893 5569 13918 8958 12137
1333 4725 4458 5905 17166 13851 4458 17 ... 6706
13850 12224 6929 14991 7382 15857 67893
5870 17553 67893 5870 5454 16102 15217
21001 17388 7446 23638 18222 6719 14331
3156 23552 22096 21604 4797 9497 22464
23610 18140 22260 5905 13347 20420 39689
6929 5275 18507 52262 1340 22049 13339 11265 22295
10439 14814 4178 6992 8784 7632 7357 6926 52262 11267
21100 21272 9346 9559 22464 15874 18502 18500 15857
2188 5376 7381 98092 16127 13486 9350 9220 76036 14219
5144 2831 17920 11347 17142 11264 7667 7762 15099 9110
10482 97556 3569 3670

BERNSTORFF.

Charge German Embassy.

The Zimmermann Telegram was sent from Germany to Mexico in coded form.

In February and March, German submarines sank several US ships. Britain also informed the United States about the Zimmermann Telegram. Mexico remained neutral, but the US public was outraged when news of the telegram was released. President Wilson asked Congress to declare war on Germany, and it did so on April 6, 1917. The United States created the American Expeditionary Forces (AEF) to fight in Europe. General John J. Pershing led the AEF.

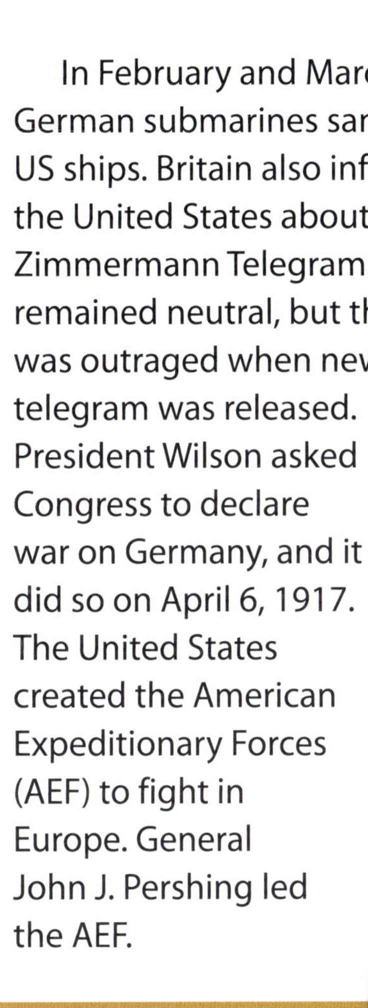

General John J. Pershing, who led US forces in World War I, had served in the American Indian Wars early in his career.

As the United States was entering the war, the Russians were leaving it. In March 1917, Tsar Nicholas II was overthrown during the Russian Revolution. Vladimir Lenin came to power. Lenin wanted to end the war. He gave control of Poland, the Baltic countries, and Finland to Germany. The Germans could now turn their full attention to the west.

It took time to train, supply, and transport US troops. US forces finally arrived on the battlefields of Europe in the spring of 1918. In March 1918, Germany began the Second Battle of the Somme in France. German forces pushed forward 40 miles (64 km). But the Allies, now with US troops and using the new tanks, stopped the forward movement of the Germans.

US troops arrived on battlefields that had already seen an enormous amount of destruction in the past three years.

At the Battle of Saint-Mihiel in early September, US troops played a vital role. Approximately 500,000 US soldiers and 100,000 French soldiers fought in this battle near the border of France and Germany. Above them were about 1,500 US aircraft,

Eddie Rickenbacker was the top US pilot of the war, shooting down 26 enemy aircraft.

US troops establish a machine-gun position during the Meuse-Argonne offensive.

representing a new frontier in US military technology. The Allied forces won the battle and began preparing for a major new offensive.

The Allies launched the Meuse-Argonne offensive in September. More than 1.2 million Americans fought in the battle, and about 26,000 of them died. The German army was driven back, and it became clear that Germany could no longer win. The two sides signed an armistice on November 11, 1918, ending the fighting.

## CONSEQUENCES

In 1919, the Treaty of Versailles officially ended the war between the Allies and Germany. The treaty forced the Germans to take blame for World War I. Germany lost territory and had to reduce its military forces. It also had to pay an enormous sum for damages to Allied countries. The treaty's harsh terms led to resentment that contributed to the formation of the Nazi Party in Germany in 1920. The Nazis were later responsible for starting World War II.

During the peace discussions after World War I, the Allies formed the League of Nations. It was the first international organization founded to promote peace and security around

The agreement ending the war was signed at the ornate Palace of Versailles in France.

The National World War I Museum and Memorial in Kansas City, Missouri, honors the sacrifices of the US soldiers who died in World War I.

the world. Sixty-three countries became part of the League of Nations during its existence. Although President Wilson was a driving force behind the creation of the organization, the United States did not join. Joining required the approval of Congress, and Republican leaders in Congress opposed the League.

## AT A GLANCE

**US Forces**: 4,355,000
**US Military Deaths**: 116,516
**Civilian Deaths**: 13 million

# WORLD WAR II (1939–1945)

World War I left Germany in ruins. The Treaty of Versailles that ended the war took German territory and forced Germany to pay for damages in the war. German citizens were angry. Adolf Hitler took advantage of that anger. Starting in 1921, he led the National Socialist German Worker's Party, or Nazi Party. He blamed groups such as Jews and communists for the nation's problems. In 1933, Hitler was appointed the

Hitler, *top right*, formed a cult of personality around himself, forcing Germans to give him a Nazi salute at public events.

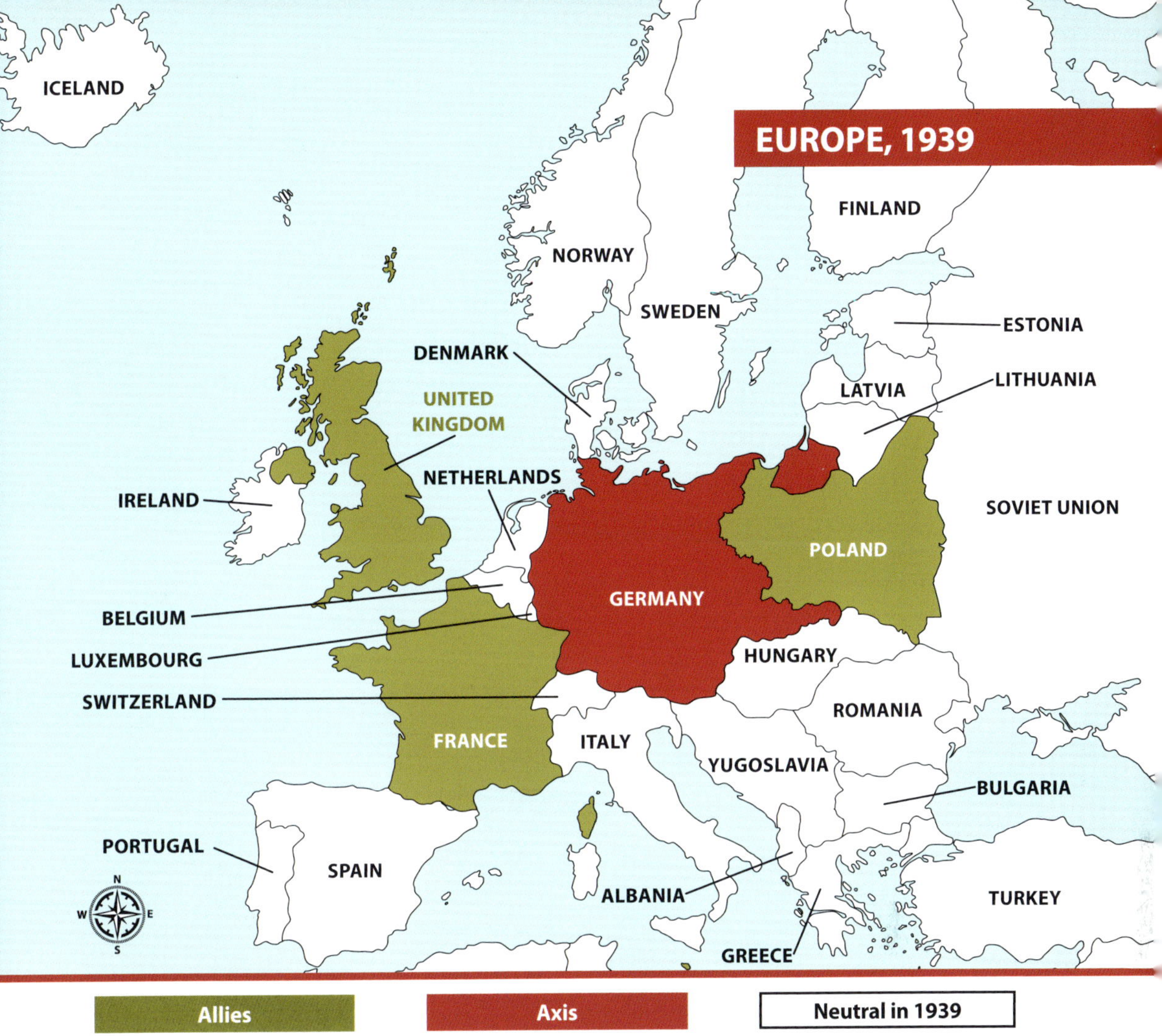


chancellor of Germany. The next year, he declared himself the Führer, or leader, possessing absolute power over the country.

Hitler rebuilt Germany's military and began expanding the country's territory. In March 1938, Hitler annexed Austria. He then turned to seizing part of Czechoslovakia. UK prime minister Neville Chamberlain did not want war with Germany. In the Munich Agreement, he convinced the Czechs to give Hitler what he wanted. This strategy became known as appeasement. It did not stop Hitler's desire for expansion.

## WAR IN EUROPE BEGINS

When the Polish government refused to cede land to Germany, the United Kingdom announced it would protect Poland in case of attack. On September 1, 1939, Germany invaded western Poland, starting World War II. The United Kingdom and France then declared war on Germany.

Germany defeated Poland in only 35 days. This encouraged Hitler to continue his plans. In April 1940, Germany invaded Norway and Denmark. In May, it

Moving quickly in tanks and trucks, and supported by aircraft, the German military swept through Poland and several other European countries.

The German bombing campaign brought stark devastation to London, the British capital.

invaded the Netherlands, Belgium, and France. The Netherlands and Belgium surrendered within a few weeks. France was defeated by June.

The United Kingdom was the only major power left fighting the Nazis in Europe. Starting in September 1940, the German air force dropped bombs on London and other cities in an attack known as the Blitz. The goal was to prepare for an invasion of the United Kingdom, code-named Operation Sea Lion.

British pilots used fighter aircraft such as the Supermarine Spitfire to defend their country from German bomber attacks.

The United Kingdom, however, was prepared. It used radar technology to detect incoming aircraft and skilled fighter pilots to defend its cities. The Blitz continued for eight months, killing 43,000 British civilians. Under the steady leadership of Prime Minister Winston Churchill, the British held on. Hitler canceled plans for Operation Sea Lion.

Hitler's major ally in Europe was Italy, led by dictator Benito Mussolini. Mussolini envisioned a new Roman Empire. In September 1940, Germany, Italy, and Japan signed the Tripartite Pact, creating an alliance in the ongoing war. The three nations became known as the Axis Powers.

At first, the United States did not get involved in the war. However, its leaders watched the events in Europe with alarm. The United States began providing the United Kingdom with

military equipment, including warships and weapons. In early 1941, Congress passed the Lend-Lease Act, which gave the president more power to provide equipment and supplies to friendly nations. US aircraft, tanks, and guns poured into the fight against Nazi Germany.

US bomber aircraft were among the items shipped overseas to friendly powers.

The German invasion of the Soviet Union made quick progress in the summer months, but eventually the Soviets were able to push back.

# THE EASTERN FRONT

Just before the invasion of Poland in 1939, Germany and the Soviet Union had signed a nonaggression pact. This agreement said the two nations would not attack each other. The Soviets then seized eastern Poland after Germany's invasion. But in the summer of 1941, Hitler prepared to break the agreement.

In June 1941, he launched Operation Barbarossa, a full-scale invasion of the Soviet Union. German forces moved rapidly through Soviet territory. Hitler expected to defeat the Soviets within two months. However, the Soviet defenses were able to prevent the Germans from taking the capital, Moscow, and Germany faced fighting through the brutal Russian winter.

Frustrated with his generals, Hitler took more direct control of the invading army.

By this time, Germany controlled almost all of Europe. The Nazis created concentration camps to hold the people the party hated, including Jews, communists, intellectuals, and others. Many of these people were subject to forced labor, and large numbers died under the brutal conditions. The Nazis also built extermination camps designed only for killing people. This mass killing is called the Holocaust. The Nazi regime killed approximately six million Jews and millions of other people.

German soldiers in occupied Poland prepare to execute a Jewish man after forcing him to dig his own grave.

# JAPAN ATTACKS

Many people in the United States, called isolationists, did not want to get involved in another European war. Laws passed in the 1930s known as the Neutrality Acts made it illegal for Americans to sell war materials to warring nations. But after Germany invaded Poland, Congress passed the Neutrality Act of 1939. It allowed warring countries to purchase arms if they paid cash and transported the weapons on foreign ships.

Within a few years, the United States would be brought into the war directly. Throughout the 1930s, Japan had been expanding its control in the Pacific. It invaded China and controlled large portions of Chinese territory. By 1941, Japan wasn't directly involved in the European war. But its Pacific expansion alarmed the United States. The United States

A Japanese pilot took a photo of Pearl Harbor just as the attack was beginning, showing the large number of vulnerable ships at the base.

blocked shipments of oil to Japan and demanded that the country leave China. Japan prepared a risky move that it hoped would prevent US interference in the nation's Pacific empire.

On December 7, 1941, Japanese planes attacked the United States' naval base at Pearl Harbor, Hawaii. It was a Sunday morning, and the base was unprepared for an attack. Waves of Japanese airplanes flew from aircraft carriers and dropped bombs and torpedoes. The Japanese wanted to destroy US naval power in the Pacific. The battleship USS *Arizona* exploded, and ships including USS *Oklahoma*, USS *West Virginia*, and USS *California* sank in the shallow harbor. The attackers also destroyed more than 180 aircraft on the ground. More than 2,300 Americans died in the surprise attack, which lasted just over one hour.

The Manzanar Relocation Center, one of the places where Japanese Americans were held, was located in the California desert.

At the same time, Japan rapidly invaded many areas around Asia, including the Philippines. Congress declared war on Japan on December 8. Within days, Germany and Italy declared war on the United States. The United States was now fully involved in a global conflict.

The US government worried that Americans of Japanese ancestry might side with Japan. President Franklin D. Roosevelt signed Executive Order 9066, which gave the military the power to forcibly relocate people considered a possible threat to national security. As a result, the US government ordered approximately 120,000 Japanese Americans to leave their homes and live in incarceration camps during the war.

These people lost their homes, businesses, and personal belongings.

## THE TIDE TURNS IN EUROPE

Following a major military buildup, the United States was ready to begin fighting in the European war by late 1942. In a plan called Operation Torch, designed to secure control of the Mediterranean Sea, US troops landed in North Africa in November. US and British forces faced challenging combat, but by May 1943 the Axis forces in the region surrendered. The Allies invaded Italy in the summer of 1943, and Mussolini was removed from power by July.

US forces used the M3 tank in North African clashes such as the Battle of the Kasserine Pass.

Meanwhile, the struggle between Germany and the Soviet Union continued. A German attack on the Soviet city of Stalingrad that began in the summer of 1942 continued into early 1943. The Soviet troops successfully defended their city. The cost was enormous, with 800,000 Axis casualties and 1.1 million Soviet casualties. The Soviet Army began steadily pushing the invaders back toward Germany.

On June 6, 1944, US, British, and Canadian troops landed on the beaches of France under the command of US general Dwight D. Eisenhower. On the day of the invasion, known as D-Day, 5,000 ships and 10,000 planes carried 156,000 troops to the European coast. Allied aircraft prevented the Germans from

The US assault on Omaha Beach in Normandy was the bloodiest portion of the D-Day invasion.

An American soldier, *left*, and a Soviet soldier celebrated when their armies linked up in Germany in the spring of 1945.

bringing in troops and supplies. The Allies advanced into France through fierce German resistance. They liberated the French capital, Paris, on August 25.

In early 1945, Allied troops approached Germany from both the east and the west. Soviet troops entered Berlin, the German capital, in April. Hitler hid in an underground bunker there with his staff, where he died by suicide on April 30. The war ended officially in Europe on May 8, 1945.

# THE WAR IN THE PACIFIC

The United States was fighting a two-front war. At the same time the military was preparing to fight in Europe, it was trying to stop the Japanese advance in the Pacific. US admiral Chester Nimitz led the fight. To win on this vast battlefield, he had to know where the Japanese would strike next. Nimitz had help from US intelligence. Commander Joseph Rochefort and his codebreakers cracked the Japanese code, allowing them to figure out the enemy's plans.

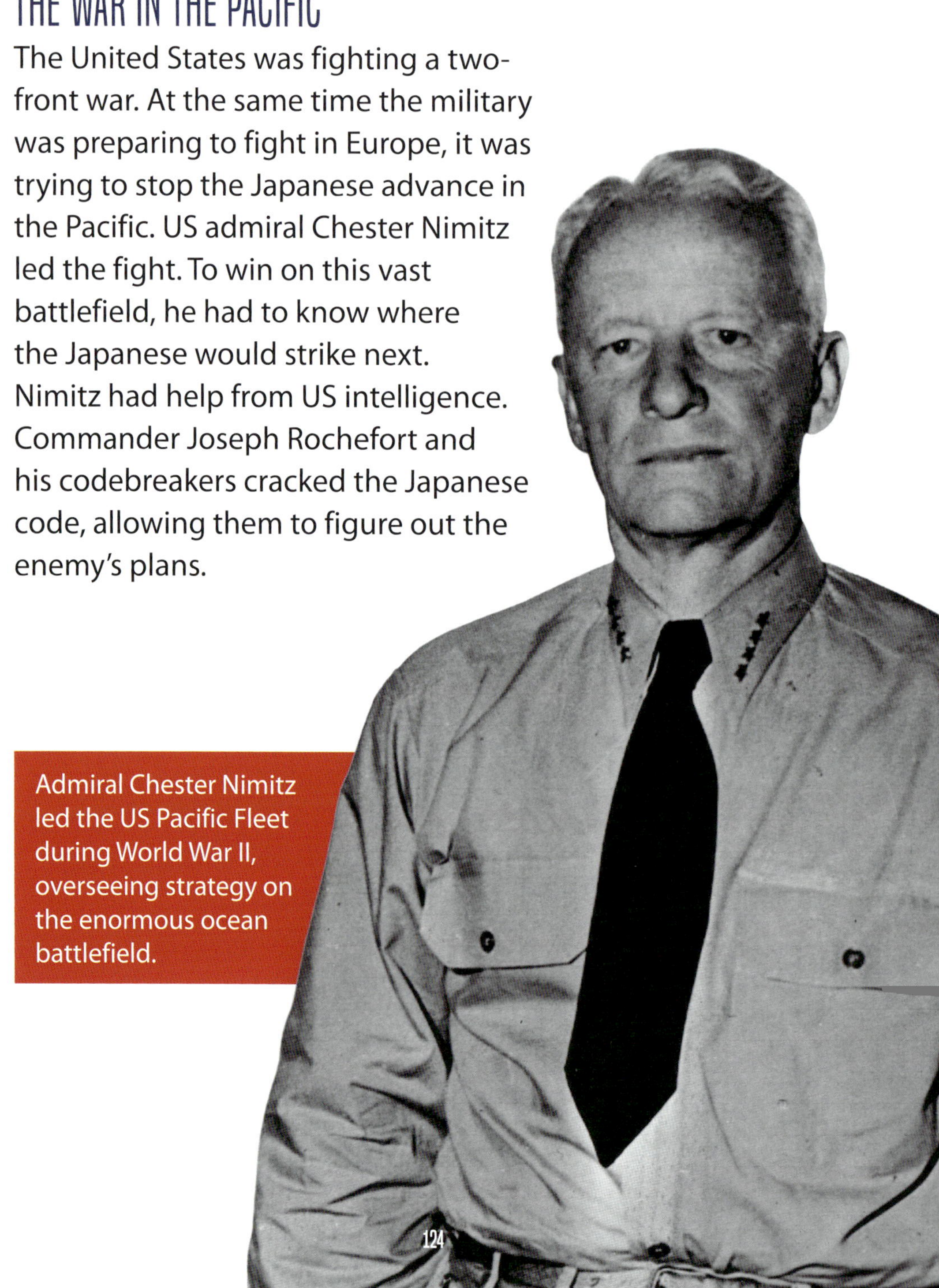

Admiral Chester Nimitz led the US Pacific Fleet during World War II, overseeing strategy on the enormous ocean battlefield.

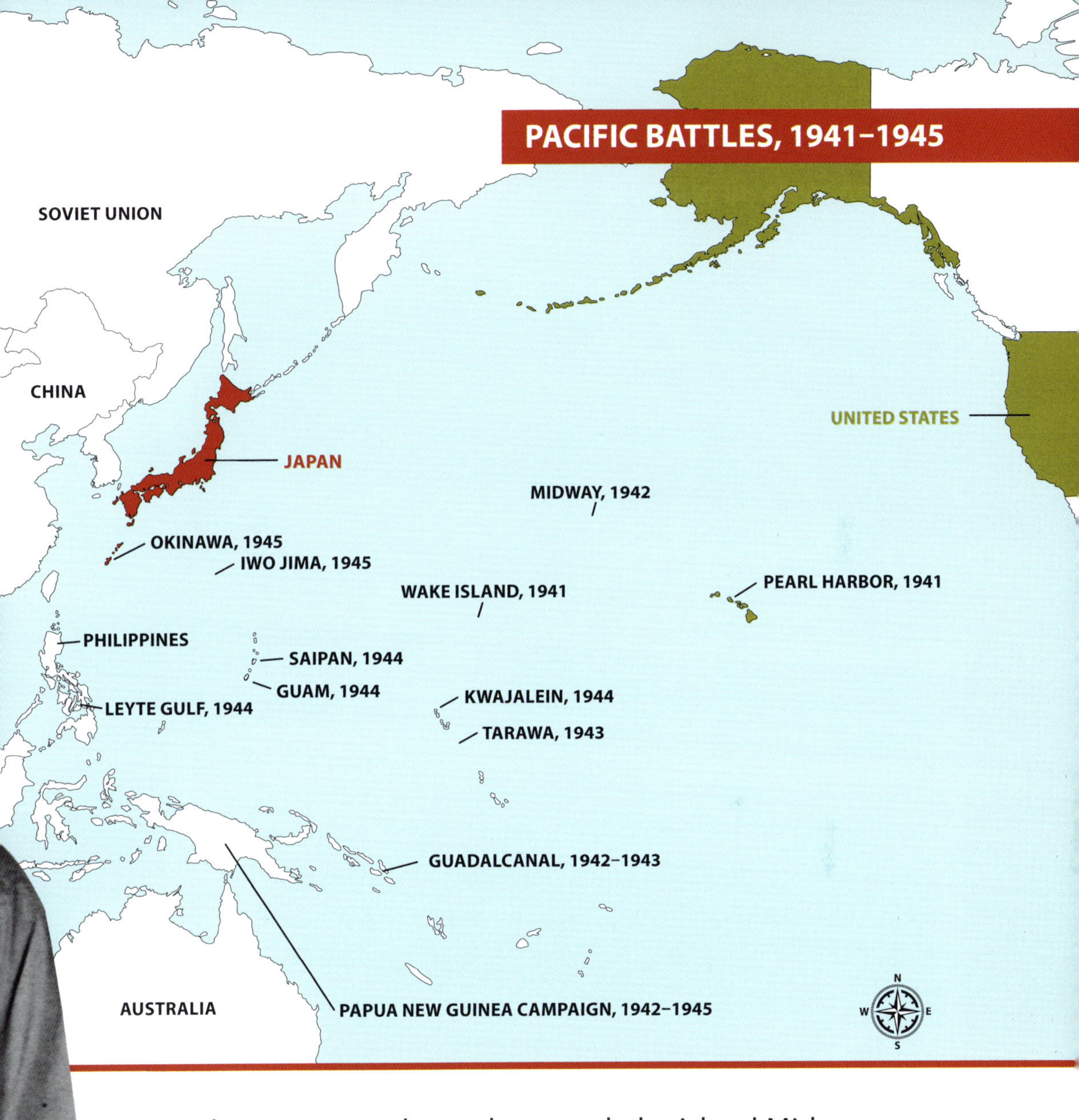


The Japanese planned to attack the island Midway, which is about halfway between Asia and North America, in June 1942. Nimitz's forces were ready. On June 4, planes from US aircraft carriers sank all four Japanese aircraft carriers present at Midway. The US victory was a key turning point in the Pacific war.

During the brutal fighting in the Pacific, US troops sometimes used flamethrowers to kill enemy troops.

On August 7, 1942, US Marines landed at the island of Guadalcanal in the western Pacific. Air, sea, and ground forces from both sides battled for six months. Finally, on February 8, 1943, the last of the Japanese forces on Guadalcanal were defeated. US troops fought bloody battles against the Japanese in Papua New Guinea, as well as on islands including Tarawa, Saipan, and Kwajalein, making progress across the Pacific. In 1944, the Allies invaded the Philippines. In early 1945, they landed at Iwo Jima and Okinawa, getting close to Japan's home islands. The Japanese fought fiercely, and casualties were high.

Meanwhile in the United States, the secret program known as the Manhattan Project was developing the world's first atomic bomb. Following a successful test in July 1945, the United States prepared to use the weapon against Japan. On August 6, 1945, a US plane dropped an atomic bomb on the city of Hiroshima. Approximately 70,000 people died instantly. Tens of thousands later died of radiation poisoning. Three days later, a plane dropped a second atomic bomb on Nagasaki. Another 40,000 people died instantly. Japan formally surrendered on September 2, ending the war.

The atomic bomb reduced much of Hiroshima to rubble.

# A NEW WORLD ORDER

Before the end of the war in Europe, representatives from 50 countries gathered in San Francisco, California, in April 1945 to discuss the postwar world. They wrote and signed the UN Charter, which formed the United Nations (UN). This international body replaced the League of Nations, which had failed to prevent war after World War I.

World War II left Europe in ruins. Ground warfare and aerial bombing had destroyed cities, leaving millions of people dead and millions more without homes. In 1948, President Harry S. Truman signed the European Recovery Program, commonly

President Truman spoke at a 1945 United Nations conference in San Francisco, California.

The fortified Berlin Wall separating communist East Berlin from democratic West Berlin became a symbol of the Cold War.

known as the Marshall Plan. The plan gave financial aid to Europe for rebuilding.

After the war, the Soviet Union expanded the reach of communism into eastern Europe. Germany was divided into West Germany and East Germany. The Western Allies formed a democratic West Germany, and the Soviets formed a communist East Germany. The competition between communism and democracy became the basis for the Cold War, a power struggle between the United States and Soviet Union that would last into the early 1990s. These two nations emerged from World War II as superpowers that shaped the postwar world.

## AT A GLANCE

**US Forces**: 16 million
**US Military Deaths**: 407,318
**Civilian Deaths**: 45 million

# KOREAN WAR (1950–1953)

Kim Il-sung led Korean soldiers in the Soviet Army during World War II and later became the leader of North Korea.

The Korean War was a civil war fought between South Korea and North Korea. Japan occupied the Korean peninsula between 1910 and the end of World War II. When Japan lost the war, the United States and the Soviet Union agreed to divide responsibility for Korea. The United States

Syngman Rhee earned a PhD from Princeton University in 1910 and worked toward Korean independence during the Japanese occupation.

administered the south and the Soviet Union administered the north. The country was divided at the 38th parallel of latitude. In the north, the Soviets supported Kim Il-sung as leader of the communist Democratic People's Republic of Korea. In the south, the United States backed President Syngman Rhee in the Republic of Korea.

Although the Soviet Union and United States had been allies during World War II, their alliance became strained after the war. Dictator Joseph Stalin led the Soviet Union. He had spread communist governments into the countries of Eastern Europe. The countries of Western Europe feared that communism would spread to democracies in the West.

Both the United States and Soviet Union had arsenals of nuclear weapons. Any direct fighting between them could result in global devastation. Instead, they struggled for

dominance in economics, political influence, and technological development. This conflict became known as the Cold War.

In North Korea, Kim grew his army. Many North Koreans fled to South Korea. In South Korea, the United States withdrew its military forces in the late 1940s, leaving South Korea to run itself. Kim wanted to unite the two Koreas under his control. With approval from Stalin, Kim invaded South Korea in June 1950. Kim's forces quickly took the capital, Seoul. However, they failed to prevent the South Korean army from escaping.

South Korean troops head toward the front to fight the North Korean invasion.

# INTERNATIONAL HELP

President Truman went to the UN to ask for assistance for South Korea. The UN passed Security Council Resolution 82, which called for North Korean forces to withdraw. North Korea

The UN discussed responses to the Korean situation in the summer of 1950.

MacArthur, *center*, played an important role in World War II and later led US forces in the Korean War.

ignored the resolution. Two days later, Resolution 83 asked UN members to provide help to South Korea.

On July 7, 1950, the UN authorized the United States to lead a group of countries to help South Korea. US general Douglas MacArthur headed the United Nations Command (UNC). These forces established a perimeter around the city of Busan on the southeast edge of the peninsula. Their goal was to prevent North Korea from capturing the rest of South Korea.

In September, General Edward M. Almond led the X Corps, a group of US Marines, Army forces, and South Korean units, on an amphibious landing in Incheon. This city is on the western coast of South Korea, near Seoul. The daring landing behind enemy lines quickly changed the direction of the war. The next day, UNC forces recaptured Seoul. General MacArthur and Syngman Rhee declared the South was free. The UNC then advanced into North Korea. The UNC quickly captured the capital, Pyongyang. North Korean troops fled into the mountains.

US troops establish a machine-gun position in North Korea in late 1950.

The US landed soldiers, vehicles, and supplies at Incheon following the successful surprise invasion.

US troops on the ground were supported by airstrikes on enemy positions.

## CHINA JOINS THE FIGHT

Kim asked for military help from Chinese leader Mao Zedong, whose country was also communist. With Stalin supplying air support, Chinese forces crossed into North Korea in November 1950. They pushed back UNC forces, surrounding them at the Chosin Reservoir. The UNC troops managed to fight their way out, and from December 9 through 24, UNC forces and Korean refugees were evacuated by sea at the coastal city of Hungnam.

The Chinese army then pushed back across the 38th Parallel and recaptured Seoul. In May 1951, the UNC took the city yet again. The front lines soon became stationary around the 38th parallel. For the next two years, neither the UNC nor the communist forces could gain significant ground.

A US soldier takes aim with a rocket launcher amid a devastated battlefield.

## ENDING THE FIGHTING

When Stalin died in March 1953, the Chinese lost the support of the Soviet Union. The warring sides signed an armistice on July 27 that stopped the fighting. They agreed that both armies would pull back from the 38th parallel. The demilitarized zone, a neutral area 2.5 miles (4 km) wide, stretched across the peninsula, separating the two sides. A treaty formally ending

The crew of a US tank celebrates the end of the fighting in Korea.

US troops remain at the border alongside South Korean troops to defend against the possibility of a future North Korean invasion.

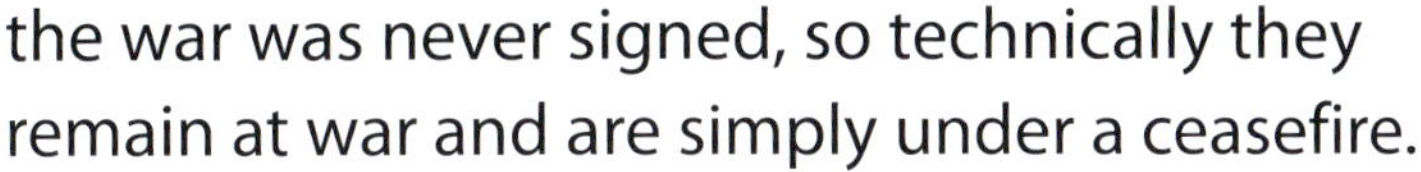

the war was never signed, so technically they remain at war and are simply under a ceasefire.

In 1953, the United States and South Korea signed a mutual security treaty. The United States agreed to protect South Korea from a northern invasion. South Korea eventually developed into a prosperous democratic country. North Korea remained an impoverished and repressive dictatorship controlled by the Kim family. Kim Jong-il took power after his father's death in 1994. When he died in 2011, his son Kim Jong-un became the leader of North Korea.

## AT A GLANCE

**US Forces**: 1,789,000
**US Military Deaths**: 36,574
**Civilian Deaths**: 1.6 million

# VIETNAM WAR (1954–1975)

Vietnam is a country in Southeast Asia. France took control of it starting in the mid-1800s, and during World War II Japan stationed its troops there. In 1941, communist Vietnamese leader Ho Chi Minh launched an independence movement. When Japan lost World War II in 1945, the communists captured the capital, Hanoi. France soon sought to regain control of Vietnam. Ho's group, the Viet Minh, fought the French in what became known as the First Indochina War (1946–1954). The United States sent aid to France, fearing the spread of communism. After years of fighting, the French agreed in 1954 to settle the dispute at a conference in Geneva, Switzerland.

In the Geneva Accords, a line was drawn at the 17th parallel of latitude. Within 300 days, all Viet Minh forces were

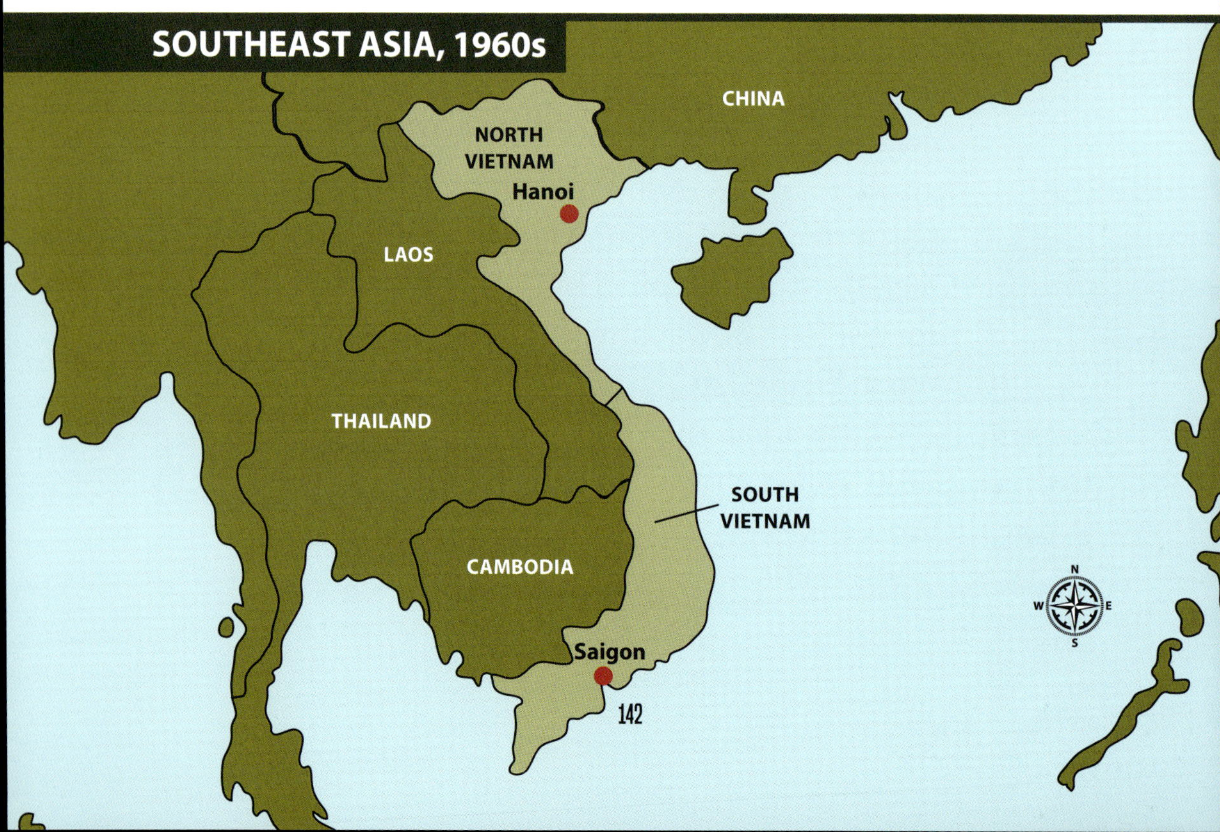

North Vietnamese leader Ho Chi Minh fought against the Japanese, the French, and eventually the Americans.

to withdraw to the north. The French were to withdraw to the south. Elections would be held before July 1956 to reunify Vietnam. Because North Vietnam controlled most of the country, the United States and South Vietnam did not agree to the terms. South Vietnam refused to hold joint elections with North Vietnam in 1956.

Ngo Dinh Diem led South Vietnam from 1954 until his death in 1963.

The two Vietnams separated. China and the Soviet Union supported a communist government in North Vietnam. In South Vietnam, President Ngo Dinh Diem, supported by the United States, built up the military. He was a Roman Catholic and suppressed the majority Buddhist population. This made him unpopular.

By 1963, a group of communist guerrillas called the Vietcong was gaining power in South Vietnam. The United States stopped supporting Diem, and he was killed in a military coup. By this time, the United States had 17,000 troops acting as advisers in South Vietnam to train South Vietnamese troops.

US advisers trained South Vietnamese soldiers in the use of weapons such as mortars.

## US FORCES DEPLOYED

In August 1964, President Lyndon Johnson informed the American people that a US ship, the USS *Maddox*, had been attacked in the Gulf of Tonkin near North Vietnam. Johnson asked Congress for the authority to defend against further attacks. In response, the US Congress passed the Gulf of Tonkin Resolution on August 7, 1964. Under the US Constitution, only Congress has the power to formally declare war. The Tonkin Gulf Resolution gave Johnson a way to get around this limitation.

Johnson ordered US Air Force bombers to strike North Vietnam starting in February 1965. The United States then started sending combat troops. By July of that year, 75,000 US soldiers were

President Johnson signs the Gulf of Tonkin Resolution, giving him authority to deploy US forces to Vietnam.

Helicopters were crucial to the US war effort in Vietnam, allowing the military to move troops and equipment to remote areas.

serving in Vietnam. The US troops had superior weapons and technology. However, the communist soldiers used guerrilla tactics, launching attacks on US forces and then disappearing into the jungle. There were few traditional large-scale battles like the ones that had defined past wars. The Vietcong and North Vietnamese lost many more soldiers than the United States. Yet US losses continued to mount with little sign of progress in 1966 and 1967.

A B-52 bomber of the US Air Force drops a load of 750-pound (340 kg) bombs over Vietnam.

A large-scale bombing campaign by the US Air Force also seemed to have little effect. By the start of 1967, US planes had dropped more bombs on North Vietnam than were used against Japan during all of World War II. Economic support from the Soviet Union and China meant that destroying North Vietnamese factories and military facilities was ineffective. Those countries also provided North Vietnam with modern air defenses, including radar equipment and anti-aircraft missiles. Hundreds of attacking US aircraft were shot down.

In January 1968, the war took a dramatic turn. The communists carried out a major attack on the Vietnamese Lunar New Year holiday, Tet. The attack became known as the Tet Offensive. Vietcong and North Vietnamese troops attacked more than 100 cities and military bases within a few weeks. However, the communists took very heavy casualties, and US and South Vietnamese forces soon regained control of these areas. Even so, the large-scale attack showed the US public that, despite the claims of the US government, victory in the Vietnam War did not seem to be near.

The Tet Offensive did not lead to lasting gains for the communist forces, but it suggested the war was far from over.

## OPPOSITION TO WAR

From the beginning, many people in the United States protested the country's involvement in Vietnam. Some people didn't approve of Diem's repressive government. Others protested involvement in a civil war. Many demonstrations took place on university campuses.

In addition, students began protests against Dow Chemical. The company produced napalm, a chemical that melted the

US forces sprayed chemicals on Vietnamese forests to eliminate tree cover where enemy troops could hide. These chemicals had harmful health effects on the Vietnamese people as well as the US soldiers who handled the chemicals.

Reporting by respected journalist Walter Cronkite, *left*, helped shift public opinion about the Vietnam War.

flesh when sprayed on people. The military used napalm bombs in Vietnam.

On February 27, 1968, Walter Cronkite, a trusted television news anchor, declared that he thought the war could not be won. Many viewers found his statement persuasive. A few weeks later, President Johnson announced that he would not run for a second term as president, in part because of the unpopularity of the war.

Incidents of US troops involved in war crimes also led to further protests. One of the most infamous incidents became known as the My Lai Massacre. On March 16, 1968, US troops attacked a stronghold of the Vietcong at a village called My Lai. As the villagers fled, US troops burned homes and massacred men, women, and children. Despite attempts to cover up the crimes, US Army investigations found that more than 500 civilians were murdered. No US officers were found guilty of covering up the crimes. One officer, Lieutenant William Calley, was found guilty of premeditated murder.

# THE UNITED STATES WITHDRAWS

Richard Nixon was elected president in 1968 with a promise to end the war. Still, it continued after he took office. Two years later, he ordered attacks on enemy supply lines in the neighboring countries of Laos and Cambodia. War protests continued. On May 4, 1970, the Ohio National Guard opened fire on unarmed protesters at Kent State University. Four students were killed and nine were injured.

The Kent State shootings became one of the most infamous moments in the Vietnam War protest movement.

People desperately evacuated from Saigon as North Vietnamese forces closed in on the city.

The United States, North Vietnam, South Vietnam, and the Vietcong signed the Paris Agreement in January 1973 to end the war. The United States agreed to withdraw forces within 60 days. The last US troops left Vietnam on March 29, 1973.

Although the war was over for the United States, the North launched an attack on the South Vietnamese capital, Saigon, in early 1975. On April 20, the North overran the government. Nine days later, remaining US personnel at the American embassy evacuated Saigon. Many South Vietnamese people fled the country, and the United States accepted large numbers of Vietnamese refugees. Vietnam became a unified communist country.

## AT A GLANCE

**US Forces**: 2,594,000
**US Military Deaths**: 58,220
**Civilian Deaths**: 2 million

# PERSIAN GULF WAR (1990–1991)

On August 2, 1990, Iraq invaded Kuwait, a small nation to its southeast. Iraq had recently fought the Iran-Iraq War (1980–1988). The country had large debts. Many of them were owed to neighboring countries, such as the United Arab Emirates (UAE) and Kuwait. When Kuwait refused to forgive the debts, Iraqi president Saddam Hussein ordered his troops to invade the country. He justified his aggression toward Kuwait by stating that Kuwait had been part of Iraq in the past. Kuwait

Saddam Hussein took power in Iraq in 1979 and fought a long war against Iran in the 1980s.

Iraq had one of the world's largest militaries at the time of the Persian Gulf War.

has rich oil fields, and Saddam Hussein wanted to have them under Iraqi control.

Iraq's forces took control of Kuwait within a few hours. Many members of the Kuwaiti royal family fled to Saudi Arabia. Saddam Hussein announced on August 28 that Kuwait was Iraq's newest province.

## WORLD RESPONSE

Kuwait was an important oil-producing country, and US president George H. W. Bush came out strongly against the Iraqi aggression. The UN passed several resolutions demanding that Saddam Hussein withdraw his forces, imposing economic sanctions on Iraq until it complied. After Saddam Hussein ignored the resolutions, the UN passed Security Council Resolution 678, which allowed the use of force to free Kuwait.

Bush formed a coalition of 35 countries, including the Middle Eastern countries of Saudi Arabia, Syria, and Egypt. General Colin Powell, chairman of the US Joint Chiefs of Staff, advised that the force be overwhelming, gather broad international and domestic support, and have a clear exit strategy.

President George H. W. Bush, *left*, and General Powell, *right*, played key roles in the Persian Gulf War.

The first phase of the coalition plan was Operation Desert Shield. The US military positioned its soldiers, ships, and aircraft to defend Saudi Arabia, preventing a possible Iraqi invasion. Troops from the United States, the United Kingdom, Egypt, France, and other nations deployed to the area. US general Norman Schwarzkopf led the coalition forces.

The US Air Force deployed F-15E Strike Eagle aircraft to Saudi Arabia as part of the Operation Desert Shield buildup.

The stealthy F-117 Nighthawk was used extensively for precision strikes on Iraqi targets during Operation Desert Storm.

On October 30, the UN gave the Iraqis 45 days to leave, or they would be forced out. When Iraq did not withdraw by the deadline, Operation Desert Shield transformed into Operation Desert Storm. In mid-January, coalition troops began aerial bombing of Iraq. Using cutting-edge technology, including stealth aircraft, they destroyed palaces, military sites, and power plants in Iraq.

The US made heavy use of M1 Abrams tanks in its ground attacks on the Iraqi military.

On February 24, ground troops of the coalition attacked, using overwhelming firepower to push Iraqi troops out of Kuwait within 100 hours. Retreating Iraqis set fire to Kuwait's oil fields. UN Resolution 687 brought a permanent cease-fire. The UN demanded reparations from Iraq totaling $52.4 billion. This money would compensate for damages done in Kuwait during the occupation. The money was paid from Iraqi oil revenues until all claims were eventually settled in 2022.

Resolution 687 also put a trade embargo on Iraq until it agreed to destroy its weapons of mass destruction. This included nuclear, biological, and chemical weapons programs as well as all missiles that could go more than 90 miles (150 km). In addition, Iraq would have to allow inspectors into the country to ensure compliance with weapons destruction. Years later, conflicts over this inspection process would lead to another war between Iraq and the United States.

## AFTERMATH

The Persian Gulf War was brief, but it caused massive environmental destruction. The Iraqi troops destroyed more than 700 oil wells in Kuwait. Sixty million barrels of oil spilled, contaminating millions of cubic meters of soil. The wells burned for nine months, causing massive air pollution. Kuwait lost two-fifths of its freshwater aquifer due to oil contamination. Oil contaminated 932 miles (1,500 km) of coastline, and 65,000 cubic yards (50,000 cubic m) of raw sewage entered Kuwait Bay each day after the destruction of sewage facilities. US troops'

Burning oil wells poured thick, black smoke into the skies over Kuwait.

F-16 fighters were among the US jets that enforced the no-fly zones over Iraq in the 1990s.

exposure to the thick smoke from the fires resulted in health issues lasting years after deployment.

Saddam Hussein was expelled from Kuwait, but he remained the leader of Iraq. After the war, minorities in Iraq fought back against his repressive regime. Saddam Hussein's forces put down the protesters with brutality. As many as 200,000 Shiite Muslims were killed between March and October 1991. Approximately 100,000 Kurds fled to the mountains, where they were hunted down and bombed by Saddam Hussein's troops. The United States and the United Kingdom established no-fly zones over portions of Iraq to protect people from being slaughtered. The no-fly zones prohibited Iraqi aircraft in those areas and were enforced by fighter jets.

## AT A GLANCE

**US Forces**: 694,550
**US Military Deaths**: 147
**Civilian Deaths**: 100,000 to 200,000

# AFGHANISTAN WAR (2001–2021)

The US war in Afghanistan began in response to the terrorist attacks of September 11, 2001. On that day, hijackers took control of four US airplanes. Two flew into the World Trade Center's skyscrapers in New York City. The third struck the Pentagon, the US military headquarters near Washington, DC. The fourth crashed in

Osama bin Laden was quickly identified as the mastermind behind the September 11 attacks.

The September 11 terrorist attacks shocked the world and resulted in a military response against the group responsible, along with those who supported it.

Pennsylvania after passengers fought back against the hijackers. Nearly 3,000 people died in the attacks.

The terrorist attacks were committed by al-Qaeda, a militant Islamist organization. The group was founded in Afghanistan by Osama bin Laden in the late 1980s. Its purpose was to drive the Soviet Union out of Afghanistan, where the Soviets were trying to support a communist government. After the Soviet Union left in 1989, al-Qaeda began fighting against foreign influence in the Middle East. By 2001, the group was harbored in Afghanistan by the Taliban, the extremist political and religious group that ruled the country.

After the September 11 attacks, President George W. Bush declared that the United States was engaged in a Global War on Terror. This war was not against one country. It was against any person or organization that used or supported terrorism. Congress passed a resolution that gave the president the authority to use force against those who had planned the September 11 attacks. Force could also be used to prevent future attacks.

One early focus of these efforts was

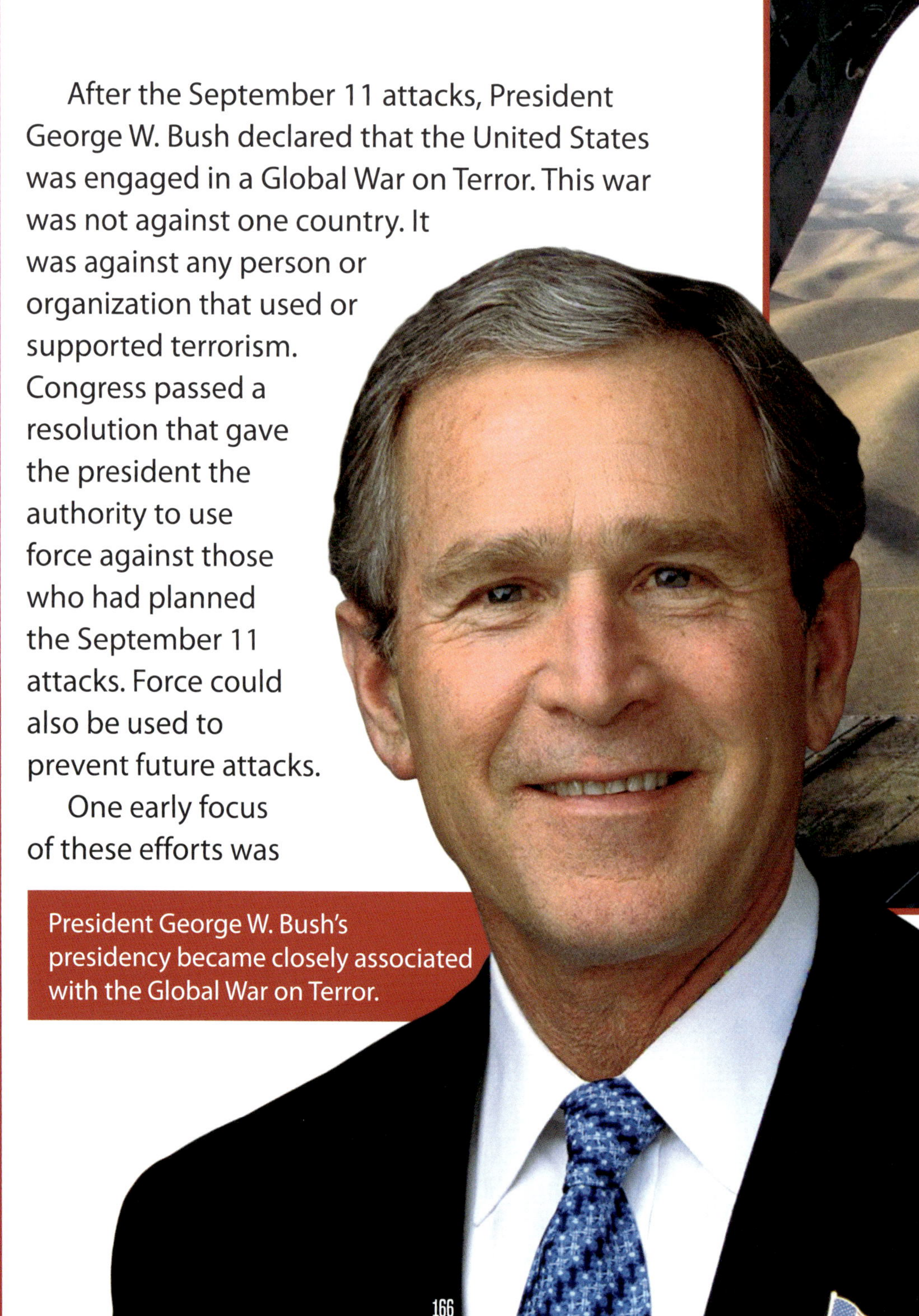

President George W. Bush's presidency became closely associated with the Global War on Terror.

US special operations troops arrived in Afghanistan weeks after the September 11 attacks.

Afghanistan. On September 26, operatives from the Central Intelligence Agency (CIA) arrived in Afghanistan to work with anti-Taliban forces in the country. Forces from the United States, the United Kingdom, and other nations provided weapons and advice to the Afghan troops and developed a plan to topple the Taliban regime.

US Marines march to a new position after taking a Taliban base in November 2001.

## OPERATION ENDURING FREEDOM

This plan, called Operation Enduring Freedom, began on October 7, 2001. The United States and the United Kingdom began bombing al-Qaeda and Taliban forces in Afghanistan. After five days of bombing, US special forces moved in. By November 2001, with help from allied troops from Australia, Canada, France, and Germany, the coalition forces toppled the Taliban government. The Taliban leader, Mohammad Omar, fled

Afghanistan. He died in 2013 in Pakistan.

The al-Qaeda fighters fled east to the mountains along the Afghanistan-Pakistan border. They hid in the Tora Bora Caves. By the time the United States arrived in this region in December, Osama bin Laden had escaped.

The United States was concerned that the Taliban would return to power unless the Afghan government was strong. The forces under Operation Enduring Freedom turned to the task of building a strong democracy in Afghanistan. They set up a transitional government to rule until a permanent government could be formed. Hamid Karzai became president of the transitional government in 2002. He was then elected president of Afghanistan in January 2004.

Hamid Karzai was president of Afghanistan from 2002 until 2014.

US and coalition forces worked to prevent the Taliban from retaking parts of the country.

The next years were difficult. Violence continued from Taliban insurgents. After the 2006 elections, suicide bombing attacks increased. Corruption within the Afghan army and police made Karzai less popular. Many Afghans criticized the United States for causing civilian casualties during its operations.

# THE SURGE

The conflict in Afghanistan was in its eighth year when Barack Obama won the 2008 US presidential election. He promised to end the war. The new strategy was to put in more troops to protect the Afghan population rather than just kill insurgents. US leaders wanted to persuade many of the Afghan militants to join with the Afghan army. Then they hoped to bring peace between the Karzai government and the Taliban. When the first surge of 17,000 US troops was not enough, an additional 30,000 troops were sent in 2010. Along with the extra troops came additional air strikes, including those carried out by remotely operated drones.

Drones controlled by military personnel on the ground became important tools in the Global War on Terror.

President Obama, *left*, and his national security team received updates on the bin Laden raid in real time.

By 2011, US intelligence agencies had located Osama bin Laden. He was hiding out in Pakistan. On May 2, 2011, President Obama ordered a raid to kill or capture bin Laden. A small team of Navy SEALs flew into Pakistan and shot the terrorist leader. Nearly a decade after the September 11 attacks, their planner had been killed.

Around this time, the US government began holding talks with the Taliban. On June 22, Obama announced that US troops in Afghanistan would withdraw by the end of 2014. In September 2014, Ashraf Ghani became president of Afghanistan. He feared the sudden withdrawal of troops would cause chaos in the country. He signed the Bilateral Security Agreement to allow some coalition forces to remain in the country. On December 28, the US combat mission ended. Still, approximately 13,000 US troops remained to support and train Afghan troops.

## FINAL WITHDRAWAL

Donald J. Trump became US president in January 2017 with a promise to bring all troops home from Afghanistan. However, by September, talks with the Taliban had broken down. On February 29, 2020, the United States and the Taliban signed a peace agreement under which the United States agreed to withdraw. The Afghan government was not part of the negotiations.

US forces began leaving the country, and by January 15, 2021, only 3,500 troops remained in Afghanistan. The new US president, Joe Biden, took office later that month. He announced he would continue

The chaotic US withdrawal from Afghanistan was widely seen as a disaster.

Taliban troops equipped with stolen US uniforms and weapons seized the Kabul airport following the final departure of US troops.

the withdrawal but extend the deadline from May 1 until August 31. As US troops left, the Taliban quickly took over most of the country. The Afghan government collapsed. By the middle of August, the Taliban seized the capital, Kabul.

As the United States withdrew, thousands of Afghans who had worked with the United States and coalition forces tried to leave the country. They feared retaliation by a brutal Taliban government. As people crowded the airport to leave, a suicide bomber killed 170 Afghan civilians and 13 US troops. The last US forces left the country on August 30. The collapse of the Afghan government to the Taliban was a shattering defeat for the United States. After almost two decades of fighting and more than $2 trillion spent, the country returned to the same state as when the United States invaded.

## AT A GLANCE

**US Forces**: 775,000
**US Military Deaths**: 2,462
**Civilian Deaths**: 70,000

# IRAQ WAR (2003–2011)

In the years that followed the Persian Gulf War, Saddam Hussein still led Iraq. The treaty that ended the conflict said Iraq had to allow inspectors into its country to

UN weapons inspection teams were met with resistance by Iraqi authorities in the mid-1990s.

Air traffic controllers aboard the carrier USS *Enterprise* helped guide the US aircraft taking part in Operation Desert Fox.

ensure it did not have weapons of mass destruction. These included biological, chemical, and nuclear weapons.

Iraq was reluctant to cooperate with inspectors. In 1998, President Bill Clinton ordered Operation Desert Fox. This operation involved bombing several Iraqi military sites. Clinton hoped to make the Iraqis cooperate. However, Saddam Hussein then refused to let inspectors into the country at all.

In Bush's 2002 State of the Union address, he said that Iraq, North Korea, and Iran formed an "axis of evil," building a case for the use of force against Iraq.

After the 2001 terrorist attacks, President George W. Bush became more concerned with Saddam Hussein's refusal to let weapons inspectors into Iraq. In November 2002, the UN passed Resolution 1441. It demanded that weapons inspectors be allowed into the country. Inspectors were let into Iraq, but Bush and British prime minister Tony Blair insisted that

evidence showed Iraq was not fully complying with the UN resolution. This intelligence about Iraq developing weapons of mass destruction eventually proved to be false. The Bush administration also suggested that Saddam Hussein had links to al-Qaeda, the group behind the September 11 attacks. In late 2002, Congress authorized Bush to launch an invasion of Iraq.

# SHOCK AND AWE

Despite opposition from countries such as France and Germany, the United States and the United Kingdom attacked Iraq on March 20, 2003, with widespread and overwhelming airstrikes. The US military used the phrase "shock and awe"

US Marines enter one of Saddam Hussein's palaces in Baghdad in April 2003.

to describe this air attack, part of a strategy to stun the Iraqi defenders and sap their will to fight. A massive ground invasion followed. US and British leadership hoped to quickly topple the Iraqi regime and locate weapons of mass destruction. Many Iraqi units offered little resistance, and by April 9, US forces captured Baghdad, the capital of Iraq. Around the same time,

British troops captured the city of Basra. In northern Iraq, US paratroopers worked with Kurdish fighters to seize the cities of Kirkuk and Mosul.

On May 1, Bush landed on the aircraft carrier USS *Abraham Lincoln* off the coast of San Diego, California, riding in the back seat of a Navy jet. He arrived to give a speech about the war to the gathered sailors. Bush declared that major combat operations in the war had come to an end. Behind him was a large banner reading "Mission Accomplished." Still, Saddam Hussein and some loyal forces continued to fight. The banner was later used to criticize the administration as the conflict dragged on for several more years and US casualties mounted.

Bush, a former pilot, became the first sitting president to land on an aircraft carrier in a jet.

US troops show the tiny hole where Saddam Hussein was found hiding.

On May 23, diplomat Paul Bremer, the head of the Coalition Provisional Authority in Iraq, disbanded the Iraqi Army. This left hundreds of thousands of trained military men without jobs. Some joined insurgent groups. As violence rose in opposition to the US-led occupation of the country, US troops searched for Saddam Hussein. They killed his sons, Uday and Qusay, in a raid in June. Saddam Hussein was finally located and captured on December 14, 2003.

Meanwhile, members of Al-Qaeda in Iraq (AQI), a branch of Al-Qaeda formed after the US invasion,

continued fighting. They conducted suicide bombings. They hit holy sites in Baghdad and Karbala. In the city of Fallujah, four US contractors were brutally murdered and their bodies put on display. In April 2004, evidence was made public of torture and abuse by US troops against detainees at Abu Ghraib prison, drawing global condemnation.

News of abuse and torture at Abu Ghraib led to protests by Iraqis outside the prison.

# THE SUNNI AWAKENING

In September 2004, US troops and a rebuilt Iraqi Army battled insurgents in Fallujah. Meanwhile, Iraq worked to rebuild

Car bombs and other terrorist attacks were obstacles in restoring stability to Iraq.

Saddam Hussein had his rights read to him in an initial appearance in an Iraqi court in 2004.

its society. Elections in October 2005 gave Iraq a new constitution. Nouri al-Maliki became prime minister in 2006. In June, the leader of AQI, Abu Musab al-Zarqawi, was killed in a US airstrike.

In October 2005, Saddam Hussein was put on trial in an Iraqi court. He was convicted of genocide against the Kurds in northern Iraq in 1982. During the trial he often interrupted the proceedings, saying that the United States was really the power behind the trial. In November 2006, he was sentenced to death by hanging. The sentence was carried out in December.

President Bush replaced US secretary of defense Donald Rumsfeld with Robert Gates in November 2006 to get a new perspective on the war strategy. In January 2007, Bush sent 20,000 additional troops to Iraq and appointed General David Petraeus as the commander of US forces in the country. The United States tried a new strategy that became known as the Sunni Awakening, in which US forces formed partnerships with Iraq's Sunni Muslim population. The plan was effective. Many Sunnis who had once fought against the US occupiers changed sides to fight AQI, believing that the terrorist group was the bigger threat.

Sunni militia members joined with the United States to defend their country against AQI terrorists.

The final US Army convoy departed Iraq on December 18, 2011.

Barack Obama was elected US president in 2008 with a promise to end the war. His Operation New Dawn emphasized building up Iraqi forces and the slow withdrawal of US forces. The United States withdrew troops over the next three years. The last US troops left in December 2011.

Although the United States had removed Saddam Hussein from power and toppled his regime, it had found no weapons of mass destruction. In addition, it found no evidence that the regime had been cooperating with al-Qaeda. As a result, the credibility of US and British intelligence was greatly damaged.

## AT A GLANCE

**US Forces**: More than 1 million
**US Military Deaths**: 4,431
**Civilian Deaths**: More than 100,000

# GLOSSARY

### amphibious
Of a military invasion, launched from the sea onto the land.

### annex
To take territory from another country.

### appeasement
Giving in to a demand in the hope of preventing further disagreement.

### armistice
An agreement made by both sides during a war to cease fighting for a certain time.

### boycott
To refuse to buy goods from a business or country for political reasons.

### cavalry
Soldiers fighting on horseback.

### cede
To formally give territory to another country.

### coalition
A group of countries that agrees to work together for a cause.

### coup
The military takeover of a government.

### economic sanctions
Policies designed to punish countries by preventing them from buying or selling goods.

### guerrilla
Describing a style of combat in which a small, independent group uses nonstandard means of attack against a larger force; also, a person fighting in this style.

### morale
The mental and emotional condition of an individual or group.

### regime
The government in control of a country.

### resolution
A formal document that expresses a group's beliefs or intentions.

### terrorism
Violence against civilians in the name of a political cause.

# TO LEARN MORE

## FURTHER READINGS

Harris, Duchess, JD, PhD, and Marne Ventura. *Japanese American Imprisonment during World War II*. Abdo, 2020.

Parker, Philip. *The Civil War Visual Encyclopedia*. DK, 2021.

Ringstad, Arnold. *The Military Weapons Encyclopedia*. Abdo, 2024.

## ONLINE RESOURCES

To learn more about US wars, please visit **abdobooklinks.com** or scan this QR code. These links are routinely monitored and updated to provide the most current information available.

# INDEX

# PHOTO CREDITS

Cover Photos: Bettmann/Getty Images, front (soldiers in gas masks, Vietnam soldiers); Allan Tannenbaum/Archive Photos/Getty Images, front (tank); Library of Congress, front (Uncle Sam poster); US Army, front (people riding horses); Everett Collection/Shutterstock Images, front (dive bombers, boat exploding, Storming Fort Wagner); US Air Force, front (F-15E Strike Eagles); Emanuel Leutze/The Met, front (Washington Crossing the Delaware); Bill Morson/Shutterstock Images, back (left); Shutterstock Images, back (right)

Interior Photos: US Air Force, 1, 158, 172 (top), 172 (bottom), 185; Emanuel Leutze/The Met, 2–3; Henry Davenport Northrop/Internet Archive, 6; Smithsonian National Postal Museum, 7; Shutterstock Images, 8, 14, 16, 81, 88–89, 109; The Print Collector/Hulton Archive/Getty Images, 9, 64–65; Paul Revere/Boston Public Library, 10; Paul Revere Jr./The Met, 11; Library of Congress, 12, 30–31, 32, 37, 39, 49, 57, 60, 60–61, 66, 68–69, 71, 72, 76, 80, 83, 87, 103; Jon Bilous/Shutterstock Images, 13; Jim Madigan/Shutterstock Images, 15; Warasit Phothisuk/Shutterstock Images, 17; Everett Collection/Shutterstock Images, 18–19, 20, 31, 33, 41, 56, 59, 110, 112, 112–113; US Army, 21, 23, 121, 136, 141, 171, 184–185, 187; John Trumbull/US Architect of the Capitol, 22; MPI/Archive Photos/Getty Images, 24, 26, 38, 47, 50, 76–77; Carol M. Highsmith/Buyenlarge/Archive Photos/Getty Images, 25; Francis G. Mayer/Corbis Historical/VCG/Getty Images, 27; Stock Montage/Archive Photos/Getty Images, 28, 28–29; Picture History/Newscom, 34–35; Alfred Jacob Miller/Walters Art Museum, 36; Kean Collection/Archive Photos/Getty Images, 40–41; Frederic Remington/Amon Carter Museum, 42–43; National Archives, 44, 52–53, 86, 102, 106, 115, 120–121, 122–123, 123, 135, 146, 150; BC Images/Shutterstock Images, 45; Lindsay Jubeck/Shutterstock Images, 46–47; Library of Congress/Interim Archives/Archive Photos/Getty Images, 48–49; Wikimedia Commons, 51, 90, 92–93, 97, 104–105; Bettmann/Getty Images, 54–55, 69, 82, 88, 94, 98–99, 101, 128–129, 131, 144–145, 153; Hulton Archive/Getty Images, 55, 100–101, 108, 124–125; Corbis Historical/Getty Images, 58; Fotosearch/Archive Photos/Getty Images, 62; VCG Wilson/Fine Art/Corbis Historical/Getty Images, 63; H. Armstrong Roberts/ClassicStock/Archive Photos/Getty Images, 67; Malachi Jacobs/Shutterstock Images, 70; *Harpers Weekly*/Internet Archive, 73; US Navy, 74–75, 75, 118–119, 137, 160–161; Buyenlarge/Archive Photos/Getty Images, 78–79; Interim Archives/Archive Photos/Getty Images, 84–85; Stefano Bianchetti/Corbis Historical/Getty Images, 85; Red Line Editorial, 91, 111, 125, 142, 156, 169; Frank Hurley/State Library of New South Wales, 95; Hulton-Deutsch Collection/Corbis Historical/Getty Images, 96, 134; The Print Collector/Heritage Images/Hulton Archive/Getty Images, 99; US Army Signal Corps/American Stock/Archive Photos/Getty Images, 107; Fox Photos/Hulton Archive/Getty Images, 114–115; Jager/PhotoQuest/Archive Photos/Getty Images, 116; Laski Diffusion/Hulton Archive/Getty Images, 117; USMC/Interim Archives/Archive Photos/Getty Images, 126; Universal History Archive/Universal Images Group/Getty Images, 126–127, 130, 152; Burton/Daily Herald/Mirrorpix/Getty Images, 129; PA Images/Getty Images, 132–133; US Marine Corps, 138, 168–169, 180; Harold M. Lambert/Archive Photos/Getty Images, 138–139; UPI/Bettmann Archive/Getty Images, 140–141; AP Images, 143, 164; Horst Faas/AP Images, 144; Wally McNamee/Corbis Historical/Getty Images, 147; USAF/AP Images, 148; Tim Page/Corbis Historical/Getty Images, 148–149; Pictures from History/Universal Images Group/Getty Images, 151; Pierre Perrin/Gamma-Rapho/Getty Images, 154; Mike Nelson/AFP/Getty Images, 155; Diana Walker/The Chronicle Collection/Getty Images, 157; Scott Applewhite/AP Images, 158–159; Greg Gibson/AP Images, 162; USAF/Hulton Archive/Getty Images, 162–163; Fabina Sbina/Hugh Zareasky/Getty Images News/Getty Images, 164–165; US Government, 166, 173; Getty Images News/Getty Images, 166–167; Rahmat Gul/AP Images, 170; Aamir Qureshi/AFP/Getty Images, 174; Marcus Yam/Los Angeles Times/Getty Images, 174–175; Karim Sahib/AFP/Getty Images, 176; Michael W. Pendergrass/USN/Hulton Archive/Getty Images, 176–177; Susan Walsh/AP Images, 178–179; J. Scott Applewhite/AP Images, 180–181; Efrem Lukatsky/AP Images, 182; Khalid Mohammed/AP Images, 182–183; Scott Peterson/Getty Images News/Getty Images, 186

**ABDOBOOKS.COM**

Published by Abdo Reference, a division of ABDO, PO Box 398166, Minneapolis, Minnesota 55439. 

102023
012024

Editor: Arnold Ringstad
Series Designer: Colleen McLaren
Production Designers: Cynthia Della-Rovere and Michael J. Williams

**LIBRARY OF CONGRESS CONTROL NUMBER: 2023939635**

**PUBLISHER'S CATALOGING-IN-PUBLICATION DATA**

Names: Henzel, Cynthia Kennedy, author.
Title: The war encyclopedia / by Cynthia Kennedy Henzel
Description: Minneapolis, Minnesota: Abdo Reference, 2024 | Series: US military encyclopedias | Includes online resources and index.
Identifiers: ISBN 9781098293062 (lib. bdg.) | ISBN 9798384911005 (ebook)
Subjects: LCSH: War--Juvenile literature. | Armed conflict (War)--Juvenile literature. | Military history--Juvenile literature. | United States--Armed Forces--History--Juvenile literature. | Encyclopedias and dictionaries --Juvenile literature.
Classification: DDC 355--dc23